The Fabulous Victrola "45"

Phil Vourtsis

Photography by Bruce Willence

4880 Lower Valley Road, Atglen, PA 19310 USA

Library of Congress Cataloging-in-Publication Data

Vourtsis, Phil.
The fabulous Victrola "45" / by Phil Vourtsis ; photography by Bruce Willence.
p. cm.
ISBN 0-7643-1637-0 (pbk.)
1. Phonograph--Collectors and collecting. 2. Phonography--History. 3. Record changers--History. 4. Sound recordings--History. I. Title.
TS2301.P3 V68 2002
621.389'33'075--dc21
2002009769

Designed by Kevin Kelly
Cover by Bruce Waters
Type set in Dom Bold BT/Korinna BT

ISBN: 0-7643-1637-0
Printed in China 1 2 3 4

Published by Schiffer Publishing Ltd.
4880 Lower Valley Road
Atglen, PA 19310
Phone: (610) 593-1777; Fax: (610) 593-2002
E-mail: Schifferbk@aol.com
Please visit our web site catalog at
www.schifferbooks.com
We are always looking for people to write books on new and related subjects. If you have an idea for a book please contact us at the above address.

This book may be purchased from the publisher.
Include $3.95 for shipping.
Please try your bookstore first.
You may write for a free catalog.

In Europe, Schiffer books are distributed by
Bushwood Books
6 Marksbury Ave.
Kew Gardens
Surrey TW9 4JF England
Phone: 44 (0)20-8392-8585
Fax: 44 (0)20-8392-9876
E-mail: Bushwd@aol.com
Free postage in the UK. Europe: air mail at cost

Contents

Acknowledgments

As you can imagine, preparing the material needed for a book like this is a huge effort that requires help from subject matter experts. I would first like to thank Dr. Alexander Magoun, the Director of the David Sarnoff Library in Princeton, New Jersey, for providing detailed information about the development of the 45 rpm system and related industry developments during the period 1939 to 1950. Next I would like to thank Dan Saporito and Bob Becker for giving me access to phonographs that were not in my collection. Bob also gave me the needed final push to tackle this project. Dan was instrumental in getting me into high gear to start my collection in earnest, for he was the fellow I met in 1993 with the 45 player under his arm. Until then, I thought that I was the only person on the planet with an interest in these machines. We helped each other collect and repair them, and then branched into the paper-and-memorabilia-collecting aspect of the hobby. Thanks to Bob Havalack for providing pricing, dating, and manufacturer information, along with pictures from his collection. Thanks to Ray Tyner, Joe Centanni, and Tom Paruta for providing countless hours of hobby information both on the telephone and at the Antique Wireless Association (AWA) Meet each year in Rochester, New York. Joe and Ray also provided pictures from their collections. Thanks also to Jim Apthorpe for providing information about, and pictures of, the earliest 45 rpm records. Thank you also to Mark Schoenthal and Doug Houston for supplying pictures from their collections.

I would also like to thank the following people for their support: Chuck Rownd, Paul Childress, Charles Cummins, Bob Kolba, John Ortale, Bill Jones, Rick Weingarten, Marv Beeferman, William Bosco, John Lee, Gary Stark, Timothy C. Fabrizio, George F. Paul, and many members of the New Jersey Antique Radio Club who have provided help for this project.

Thanks to my brother, John Vourtsis, for assisting and for allowing his daughter, Suzanne Vourtsis, to pose for the cover photo.

Thanks to Dr. William Gourd for taking the time to proofread, edit the text, and provide very helpful suggestions. Thanks to Lee Wells, Bill Pauluh and Jon Butz Fascina for providing advertising information. Thanks to Jim Whartenby for inviting me into the David Sarnoff Library years ago to begin my research. Thanks to Nick Domenico for getting me involved with the Camden County Historical Society. Special thanks to previous Executive Director Paul W. Schopp and current Executive Director John Seitter of the Camden County Historical Society for their assistance in facilitating access to countless memos and advertisements from the old RCA Victor Camden facility, and, especially, access to engineer Benjamin Carson's scrapbook highlighting his career at RCA.

Thanks to Harry Duravich for helping me locate so many treasures over the years, and to Bruce Mager (Waves) for his willingness to keep an eye out for the unusual on my behalf. Special thanks to Dr. Russ Naughton of Melbourne, Australia, who was gracious enough to *give* me the April 1949 issue of *Radio Age* after winning it in an eBay auction and reading it through. That magazine was the source of all the factory photographs of the first 45 records being produced, as well as photos of the players being assembled.

Last, but most importantly, I would like to thank Bruce Willence for his top-notch professional job of photographing many of the phonographs and related items in this book. Bruce, a software developer by profession, has been an avid amateur photographer since age eleven, when he was given an old camera by his father. Over the years he has won prizes in several photo contests. He attended a Nikon School of Photography seminar and is currently enrolled in the New York Institute of Photography's Complete Course in Professional Photography, from which he learned the techniques used for the photos in this book.

Introduction

Back in the 1950s when I was a kid, there were two activities that I loved. One was playing ball on the Brooklyn streets. We played stickball, punchball, slapball, stoopball, and a hundred other creative games with a pink rubber ball. But when I was at home, my favorite activity was listening to music on our 45 rpm phonograph. Aside from enjoying the music, I used to marvel at the way the mechanism would drop each record individually, and then play it. We had a phonograph attachment that had to be plugged into a radio or television set in order to hear the music. When my parents connected it to our RCA Victor console television, I used to stand on a hassock in order to reach the phonograph. This TV had incredible sound playing through its twelve-inch speaker.

High Bounce rubber ball, more popular than baseballs or softballs to inner city kids.

In 1958, my tenth birthday present was a "New Orthophonic High Fidelity" Wood Mahogany 45 Player and the hit single "Breathless" by Jerry Lee Lewis. During the next couple of years I played so many records on it, I wore out the record changer. As the 1960s approached and stereophonic sound was introduced, I quickly abandoned my 45 player and started building component stereo systems that, of course, sounded much better. But in the early 1970s I was bitten by the nostalgia bug. Wouldn't it be nice, once more, to play those oldies on that old 45 player? That is when the quest began to refurbish my original machine. The cabinet and amplifier were still intact, but the changer had been thrown out years earlier. At an old farm sale I found the 45 rpm record changer that I needed, bought it for one dollar, and went to work. It was quite a kick to have the original player working again, and I decided that if I saw other models at flea markets or garage sales, I would pick them up.

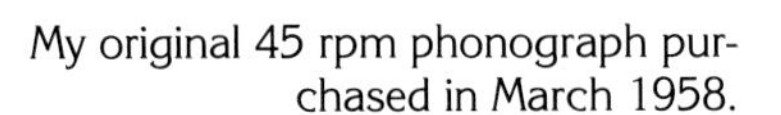

My original 45 rpm phonograph purchased in March 1958.

During the next twenty years I accumulated about ten of them. It wasn't until 1993, however, that I really became crazy about this stuff. I joined the New Jersey Antique Radio Club (NJARC), and met a fellow who also enjoyed these machines. We were at a swap meet when the President of the club told me that he had just sold one of those 45 rpm machines to some guy before I got there. I was thunderstruck. You mean there was someone else interested enough to buy one? I made it a point to find this guy and introduce myself, and we couldn't stop talking about the record players. It was at this point that I decided to widen my interest by searching for not only the machines, but also any advertisements, technical information, and related memorabilia. Over time, I built a network of contacts via the radio club and through other acquaintances and my collection has now grown to over two hundred machines, plus binders full of related information.

Another part of this hobby that I have been able to enjoy is the repair of these machines. Coming from a solid background in electrical engineering and technology, it was natural for me to want to explore the innards of these little players. Along the way I have found many technical papers that describe how to repair and adjust these mechanisms, so I consider myself an expert on such repairs. The machines are not difficult to repair, but having some mechanical aptitude is a definite advantage.

In 1995, having networked for some time with a handful of other 45 rpm phono enthusiasts in various parts of the nation, I decided to write a newsletter called *The 45 RPM Phono Gazette*. It is mailed quarterly and has been well received. I was also fortunate enough to visit the David Sarnoff Library in Princeton, New Jersey on several occasions. This allowed me access to technical papers and a great store of technical data about these machines and their history. Armed with all this information, I felt that writing a book was the next logical step. I also met the Director of the David Sarnoff Library, Dr. Alexander Magoun. To my utter amazement, I discovered that Alex had written his doctoral dissertation on "Shaping the Sound of Music: The Origins of the LP and High Fidelity, 1939-1950." This represented years of research on aspects of the very subject I wished to cover in the book, so I am gratified that Alex is providing Chapter Two: "An In-Depth Look at the Origins of the LP, the "45", and High Fidelity, 1939-1950."

You'll find that this book is sectionalized—the history of the machines is presented first, followed by detailed information about each RCA Victor table model, followed in turn by many examples of changers produced by other manufacturers, using either RCA Victor's changer or their own changer design. The final section deals with repair and restoration of the machines.

Note: Although 45 rpm records are referred to as 7 inches in diameter, they are actually 6 7/8 inches in diameter.

Price Guide Information

There are many variables that can affect the price of an item like a phonograph. Geography, type of seller, and condition are very important ones. Typically an item tends to be worth more in a metropolitan area than in a small town. Overhead has quite an impact on prices. Obviously a seller at a garage sale or church bazaar will usually be asking much less than an antique store in the mall. Prices on this type of item are higher in Florida for a different reason. Many people who reach retirement age decide to settle in Florida and get rid of their attic full of stuff before they head south, causing a shortage of such items at their new location.

Condition can bring the price down to rock bottom if it is bad enough. But rock bottom is never zero because the unit can always be sold for parts. Things like cracked Bakelite or plastic cases, missing parts, and damage from insects and rodents fall into this category. This means that you bargain hunters can still find a unit with multiple defects for $10 or $15 at a garage sale, but it can easily cost over $100 to get it in good working condition. So a benchmark must be set, otherwise prices mean nothing. Therefore, the values listed here are for phonographs that are in *excellent condition* and *working electrically* and *mechanically*. All controls are original, present, and usable. If there is a radio in the unit, it is functioning with decent reception across the dial and the dial cord is intact and working. Working does not mean, "It turns when I plug it in" or "Works great, just needs a cartridge or a tube."

For phonographs in less than excellent condition, values may be affected as follows:

• Sets that are not working can easily cost over $100 to repair

• The case can show minimal wear. If there are deep gouges in wood cabinets or cracks in plastic or Bakelite cabinets, the value can go down by 25 to 50 percent.

• If the set was poorly or improperly refinished, the value can drop by 25 to 50 percent.

• If trim is missing or dented, the value can drop by 10 to 30 percent.

• Chips or missing chunks from plastic or Bakelite sets will reduce the value by 20 to 60 percent. If the damage is not noticeable (in the back corner) the reduction will be less than if a chunk is missing in the middle of the front.

Neither the author nor the publisher accept any responsibility for losses incurred by using the pricing in this book.

Chapter One

The Way it Was

Picture yourself living in the 1870s. A relaxing evening at home would consist of reading or conversing with your family or friends in candlelight or gaslight. If you wanted to enjoy some music you would take out your guitar or violin or harmonica and play yourself a tune. Listening to a full orchestra was indeed a big event because you would have to travel to the town center or the symphony hall. With this as a backdrop you can understand how excited everyone was when Thomas Edison created the first phonograph using a cylinder in 1877. The first demonstrations of the remarkable new device, in towns and cities on the eastern seaboard, showed how the human voice could be captured and played back at any time. That was the truly striking feature of the first phonograph—the listener could, at any time, play a recording and hear the recorded voice over and over. The novelty of the recorded voice became old news within just a few years, however, and Edison lost interest in his new invention, putting it aside in favor of other activities. Only when two Bell Laboratories inventors improved on his design in the early 1880s would he go back and work on it again. Edison felt that the cylinder gave a truer reproduction of the sound than a flat disc because one did not have to deal with the distortion caused by the centripetal force that was generated by a needle tracking the grooves of a disc. Nevertheless, Emile Berliner's 1887 invention of the first practical flat disc was a decisive statement of the disc's superiority, when factors like storage and reproducibility were taken into account, and Berliner improved upon it until it became formidable competition for Edison's cylinder. Berliner found that his mixture of shellac and slate dust produced a long wearing but brittle playing surface. His formula, it turned out, would represent the composition of the disc for over fifty years to come.

To-Day's Magazine for January 1909

Edison cylinder phonograph advertisement circa 1909.

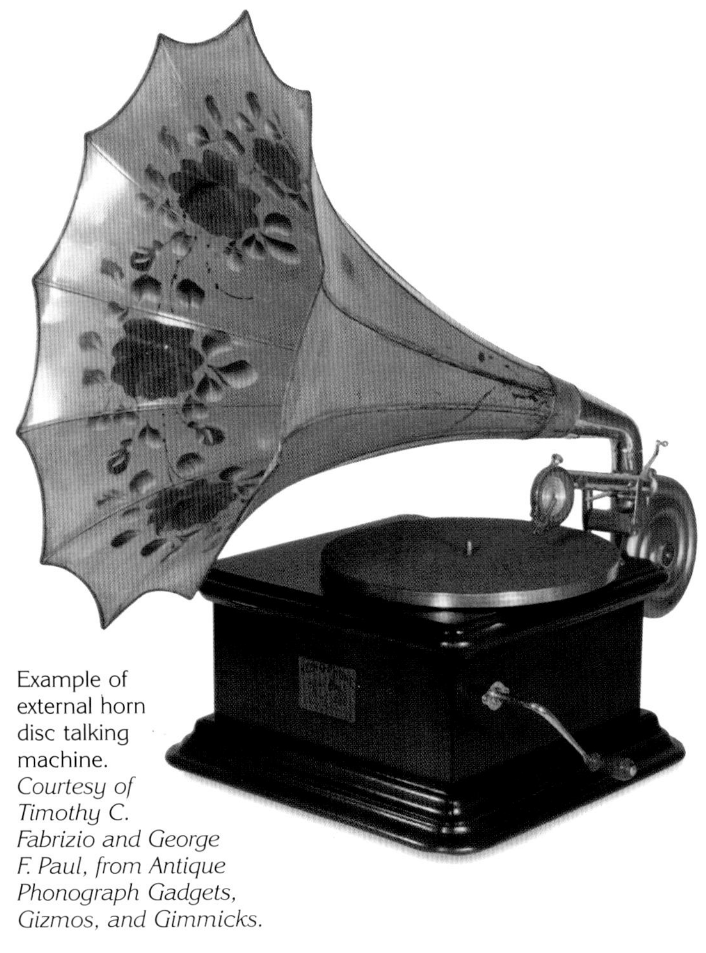

Example of external horn disc talking machine. *Courtesy of Timothy C. Fabrizio and George F. Paul, from Antique Phonograph Gadgets, Gizmos, and Gimmicks.*

The public now had a choice—to buy a cylinder playback machine or a disc playback machine. Both were completely acoustic, using no electricity. Since every acoustic machine had a horn, it was quite an interesting décor problem. At first the horn was tolerated. Later, when manufacturers dressed the horns up with colorful flower patterns, they became the focal point of the room. Then RCA Victor introduced the Orthophonic enclosure, in which the horn was hidden inside the cabinet. Once again horns were out of style.

Recording the cylinders and discs was a genuine challenge. In playback mode, both cylinder and disc machines used a stylus ("needle") to track the grooves of the recording. Vibrations of the stylus were amplified somewhat through an attached diaphragm, and the sound thereby generated was further amplified by passing it through the large horn—the precursor of today's loudspeaker. In making recordings, the process was simply reversed: the sound produced by singers and/or musical instruments was directed into the horn, thence to the vibrating diaphragm, and finally converted to vibrations of the stylus, which cut the grooves into the cylinder or the disc. The acoustic process had many inadequacies that forced orchestras to assemble in the most bizarre fashion to get the best sound balance between instruments. Loud brass instruments would be positioned in the back, at some distance from the horn, while violins and pianos would be directly in front of the horn.

Acoustic recordings of certain types of music, especially classical, have what some have referred to as a "toy town orchestra" sound. In the early days of acoustic recording there was no means of making a "master" from which copies of the recording could be mass-produced, so it was necessary to record each individual cylinder and the orchestra would have to play the music over and over. Usually up to eight cylinder machines would be operating simultaneously, recording each performance, so that each "take" would record only eight cylinders. In order to record another eight cylinders, the musicians and/or vocalists would repeat their performance. A tedious process at best, but one that created a high demand for orchestras in the early 1900s.

That all changed when Western Electric, AT&T and Bell Laboratories created the first electronic recordings using newly designed microphones and amplifiers in 1924. The change from acoustic to electronic recording, together with the advantages of the disc as its technology was improved, *e.g.* the creation of a master disc that could stamp out many shellac discs, revolutionized the industry. Sound quality was much improved, the manufacturing process was streamlined, and, most significantly, recordings could be mass-produced by the thousands, thereby reducing the price of disc records and making them available to greatly increased numbers of eager customers.

Finally, in the late 1920s, Thomas Edison, giving in to the inevitability of progress, abandoned his insistence on the utility of cylinder recordings, and began producing disc playback machines.

78 RPM Disc Reigns For More Than a Half-Century

For more than a half-century, shellac discs spinning on a turntable at 78 (more or less) revolutions per minute were the standard of the recording industry. The 78 rpm criterion was more the result of circumstance than of scientific reason. Since the early motors were spring wound, the speed had more to do with available hardware than anything else. Early record speeds were very inconsistent, and turntables rotated at a variety of speeds between 70 and 80 rpm. When the electric motor was introduced, it was possible to specify a more precise rotation criterion, and 78 rpm was decided upon. Many early disc phonographs included a speed adjustment, usually accomplished by a lever marked "slow" and "fast" to compensate for those early discs that were not recorded at exactly 78 rpm.

Attempts at Standardization (10", 12")

Early disc records were produced in a variety of diameters (7", 8", 10", 12"), but it was not long before two diameters emerged as standard: 10" discs for popular music and 12" discs for recordings of classical music. The typical running time for a popular song (3 to 5 minutes) fit nicely on the 10" disc. The longer classical selections were assigned to the 12" size. Some children's records were still produced in the 7" size.

RCA Attempts 33 RPM in Early 1930s

As recording technology continued to advance, it became clear to audio engineers and record manufacturers that if it were possible to lower the rotational speed of the record, more time would be available for putting music on that record. In the early 1930s, RCA attempted to introduce a long-playing record with a speed of 33.33 revolutions per minute, but stylus ("needle") and pickup cartridge technology were not yet up to the task. The primitive cartridges and heavy styli available to the general public wore out the new, softer vinyl records after only a few plays. This and the deep slump in sales caused by the Great Depression caused the new record to be quickly pulled from the market.

1934 saw an upturn in the record industry. Jukeboxes had been steadily growing in popularity and by 1934 they appeared all across the country; in addition, RCA Victor introduced an inexpensive 78 rpm phonograph attachment that could be plugged into any radio. RCA also introduced the "Red Label" line of superior records. Extra care was taken in choosing the finest raw materials and more attention was paid to the recordings. Sonically, most discs were steadily improved throughout the '30s and '40s with better recording techniques. But the best was yet to come, as big companies like RCA and Columbia were about to create something new and exciting.

Chapter Two

An In-Depth Look at the Origins of the LP, the "45", and High Fidelity, 1939-1950[1]

by Alexander Magoun

Overview

Between 1939 and 1950 the phonograph record underwent its greatest changes in design and capability since mechanic Eldridge Johnson innovated Emile Berliner's disc record at the turn of the century. Engineers at the corporate successor to his company, RCA Victor, and one of his long-struggling rivals, Columbia, innovated vinyl-based microgrooved records that played 45 or 33 1/3 rpm, far slower than the 78.26 rpm that had become standard since the electrification of turntable motors.[2] Researchers had been aware of the technology and the techniques necessary to make a quieter, longer-playing, less distorted record since the late 1930s, but it took over ten years for inventors, producers, and consumers to agree on the utility and shape of the new formats.

If events had proceeded as RCA Victor's engineers and marketers anticipated, there would have been only one format. Their 45-rpm Vinylite disc, 6 7/8 inches in diameter, was designed to play the vast majority of musical selections on a fast-acting record changer of limited fidelity, in response to market trends and consumer preferences. It would have been incompatible with machines designed for 78s, but RCA's management hoped to use its market power and the self-evident advantages of the new record to persuade the industry and consumers to accept a radical innovation.

The onset of World War II, the revival of Columbia as a major label, changes in RCA's organization, unprecedented expansion of the industry, fear of the appeal of other technologies, a broadening interest in high fidelity for the consumer, and the increased influence of music lovers as critics of the industry delayed innovation and changed the environment in which it would have taken place. Columbia drew on the initiative of dynamic laboratory director, Peter Goldmark, to begin research and development of a record that would play longer classical performances without interruption or significant distortion to the recorded sound. Engineers hired from other companies, including RCA, used the fundamental work published by Hunt and Pierce in the 1930s to invent a disc that accomplished that task.

The attention paid to the LP as a "revolution" or a "renaissance" in recorded sound is deserved more in retrospect than it was at the time.[3] Between the end of World War II and 1948, producers offered vinyl discs and recording techniques that provided "high fidelity" sound.[4] Nonetheless, in a period of declining sales, Columbia picked the right time to offer the new format to the market. Classical music critics, writing for a broader public, applauded the LP not for its sound quality so much as for its fidelity to the composer's intentions. David Sarnoff's pride and the need to protect the company's investment and trademarked reputation drove RCA to compete with its record designed for a different purpose. Despite the higher fidelity to the music, the 45 was criticized as another indication of the mid-culture direction RCA was taking. By 1950, the company accepted the LP with the rest of the industry while it succeeded in selling the 45 for its original purpose. These formats have remained in place as industrial standards for the phonograph record ever since, even as other technologies superceded them in the marketplace.

The Trouble with Changers at RCA Victor, 1939-1944

In 1939 RCA executives looked back at a renaissance in the phonograph industry, where record and phonograph sales increased at an average annual rate of over twenty percent for four years.[5] This was due in large part to RCA's marketing. Edward Wallerstein in the Records Department and Thomas F. Joyce in Home Instruments had promoted the production of turntables played back through radios and established the Victor Record Society to lead consumers to more sophisticated—and expensive—music and phonographs as they aged.[6]

While sales rose to pre-Depression heights, the emphasis on cheap players and the increased competition shrank profit margins. The average price of a console declined from $750 in 1928 to under $100 in 1938, when the cheapest radio-phonograph combination could be had for $49.95. RCA's unit price for phonographs dropped more than fifty percent between 1936 and 1940.[7] Efforts to add value, through better sound or new features, had not given the company the market advantage it desired.

While prices dropped, rising sales of consoles with automatic record changers indicated that consumers liked programming their own selections.[8] This demand arose in response to the ways they heard records played outside the home. Jukeboxes flourished after the end of Prohibition in December 1933. Entrepreneurs like Homer Capehart and David Rockola and their engineers provided machines with more choices and easier selection.[9] By 1939, between 225,000 and 300,000 jukeboxes absorbed thirteen to thirty million records, or up to half of all records sold.[10] At the same time, radio programmers and disc jockeys responded to the jukebox's popularity and the sales listings in *Billboard* magazine by formatting playlists, as in "Make-Believe Ballroom" and "Your Hit Parade." The syndication of these shows encouraged growing numbers of listeners to reproduce this programming at home.[11]

This trend had technological and industrial consequences. One observer noted that the rising demand for cheap record changers posed an "opportunity for further improvement in . . . record sales through standardization."[12] The 78, after all, had never been designed for changing or to formal specifications. At the end of the 1930s, engineers at different companies had developed four mechanisms for playing records consecutively, all of which had flaws. One "was the best delaminator of Columbia records ever invented," while another "systematically discriminated against Victor records, cracking them neatly."[13]

The weight of 78s on the spindle, and the different thicknesses and edges of the discs broke down the drive and separator mechanisms.[14] The drawbacks helped give changers the worst reputation among RCA Victor products.[15] Therefore, at the year-end meetings between the sales and engineering staffs, Joyce challenged engineer Ben Carson to design the perfect, fast-acting machine for the 1939 model year. Carson, who had designed all of the company's changers since Victor's first mechanism in 1927, responded that he would have to change the disc as well.[16]

The prospect of standardizing the phonograph industry to patents on players and records must have been a powerful attraction for RCA's management. 78s were brittle, heavy, and bulky. Manufacturers took losses on breakage in shipment, and as record sales rose, so too did inventory and the need for storage space by producers, middlemen, and consumers. The new system would eliminate the need to manufacture and distribute records sequenced for playback on both manual and automatic players. For RCA Victor and its licensees, fitting a record changer, radio, and the television introduced in the spring of 1939 in a console appropriate for the average living room was not feasible with current technology.[17]

To the engineers dedicated to the processing of sound, the shellac record was an insult. Noise generated by the needle abrading the blend of mineral powders and shellac meant that records still offered no better frequency-range reproduction than in 1925.[18] The Vinylite used without filler in high-fidelity commercial recordings was too expensive for standard 78s, where shellac comprised less than fourteen percent of the material.[19]

Finally, Sarnoff had to justify RCA's standing as a patent monopoly sanctioned by the United States government for quite different purposes. Without dominating the marketplace, RCA had standardized and innovated electronic components since the mid-1920s while dodging criticism and anti-trust action. Sarnoff argued that the industry benefited more than the company from its policy of patent-based innovation. New patents retained licensees' interest, demonstrated that the corporation was not resting on its monopoly, and maintained the value of a reputation and trademark that elevated sales of RCA's own products above price competition. Unwilling or unable to lead in consumer sales, Sarnoff, his marketers, and his engineers all took pride as well as profit in their position of technical leadership.[20] As the more valuable of its AM radio patents expired, the company had developed patent collections in FM radio, facsimile, and television, and was beginning to market these systems. Extending RCA's patent control to the record industry on behalf of standardization, when it had the leverage to do so, would benefit everyone concerned.[21]

Thus, management agreed with Carson. In order to protect the patent life of the system components, the project became the only secret consumer technology project in RCA's history. For the next nine years, while Carson's Advanced Development Group drew on staff from other departments and divisions as needed, lab and production space for "Project X" and the "X" record were closed to outsiders.[22]

Carson based his initial design on the intention of continuing to fulfill the slogan, "the music you want when you want it," that Victor Talking Machine Company's marketers had coined to combat broadcast radio in the 1920s. Instead of dispersing the support and changing mechanisms around the perimeter of the turntable, Carson took advantage of the design of the new disc to concentrate the parts to hold and separate the discs inside a larger spindle. The speed of the changer affected the speed of the record. After some experiments at forty and 33 1/3-rpm, Carson's team adopted forty-five rpm in 1943 as reliable for changer operation within two revolutions of the turntable, or five seconds.[23] The stability of the stack would come from smaller, lighter records. The new discs would be made of pure Vinylite, which would also eliminate surface noise. The record was also thinner than the 78, enabling the use of the expensive manmade plastic.[24]

How long should the new record play? The engineers requested a survey of the records in Victor's "Music America Loves Best" catalog. This included classical pieces divided by the composer, like symphonic movements, and did not include current popular hits.[25] Ten-inch 78s that played for about three minutes comprised most of the industry's production, but classical "Red Seal" performances on twelve-inch discs made up thirty-five percent of Victor's gross sales. Seventy percent of the classical recordings ran under five minutes, as did eighty-two percent of the MALB catalog. When all Victor records were considered by unit sales, the figure rose to ninety-six percent, and Carson assumed that the industry proportion, because of Victor's dominance of the classical market, was even higher. Using a narrower groove and one-mil stylus, Carson's group designed a disc under seven inches in diameter that could play as long as a twelve-inch 78, or five minutes and twenty seconds.[26]

Prototype of 45 rpm record player completed by Benjamin Carson's group in 1942. *Courtesy Camden County Historical Society.*

Top view of the first 45 rpm record changer in patent application. *Courtesy Camden County Historical Society.*

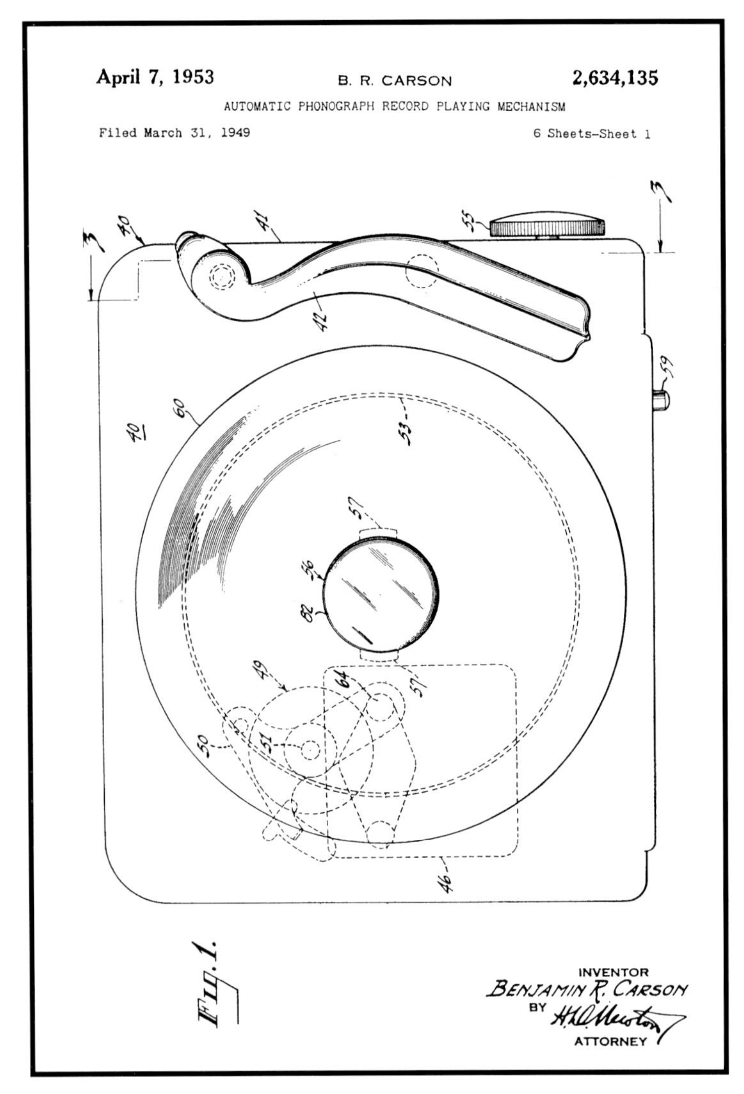

Front view of the first 45 rpm record changer in patent application. *Courtesy Camden County Historical Society.*

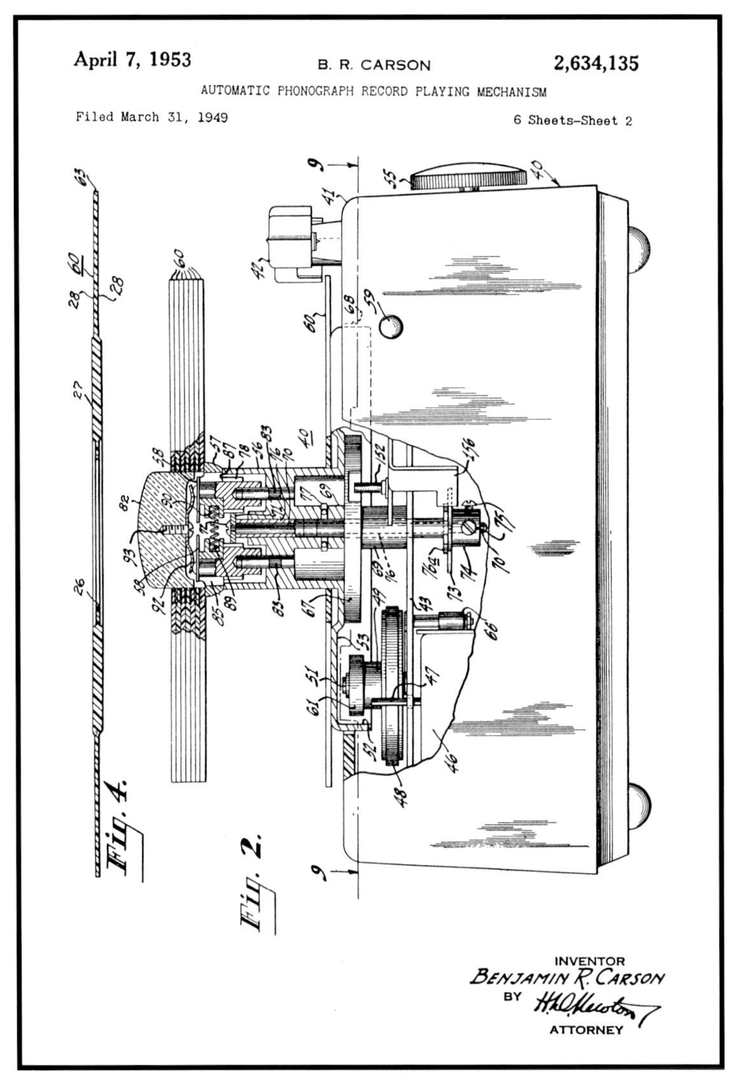

Side view of the first 45 rpm record changer in patent application. *Courtesy Camden County Historical Society.*

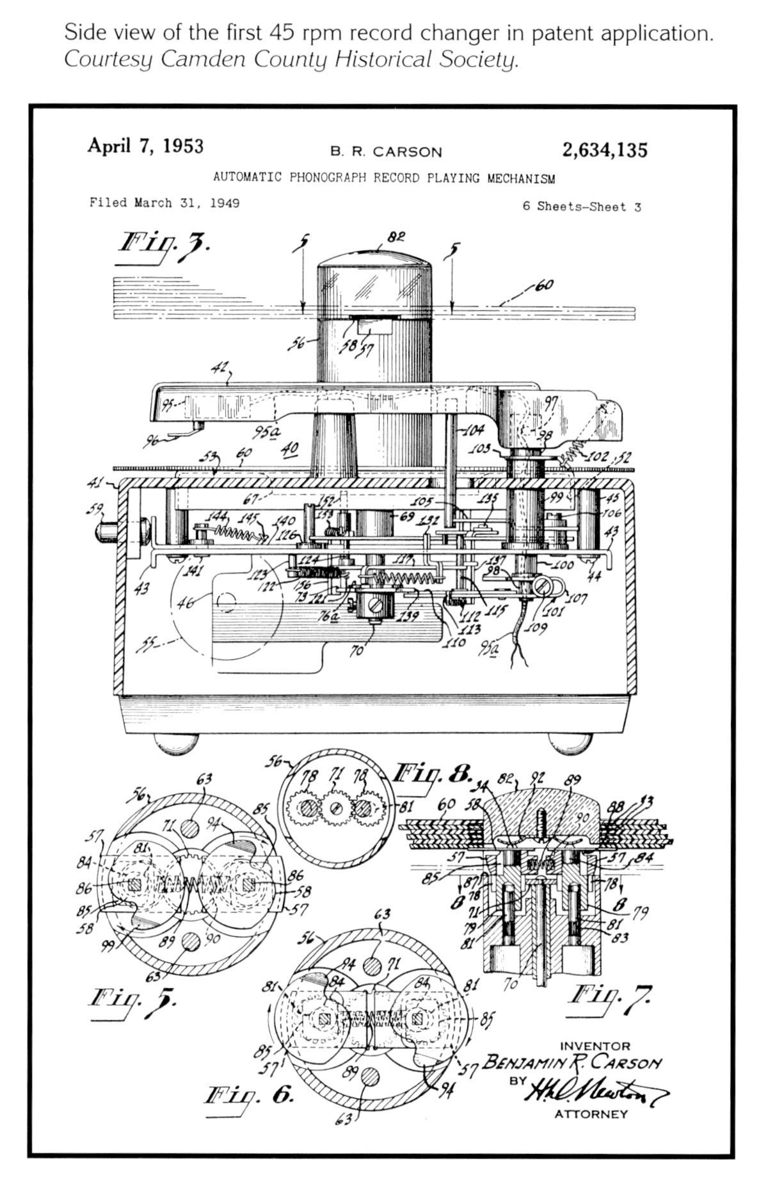

Top inside view of the first 45 rpm record changer in patent application. *Courtesy Camden County Historical Society.*

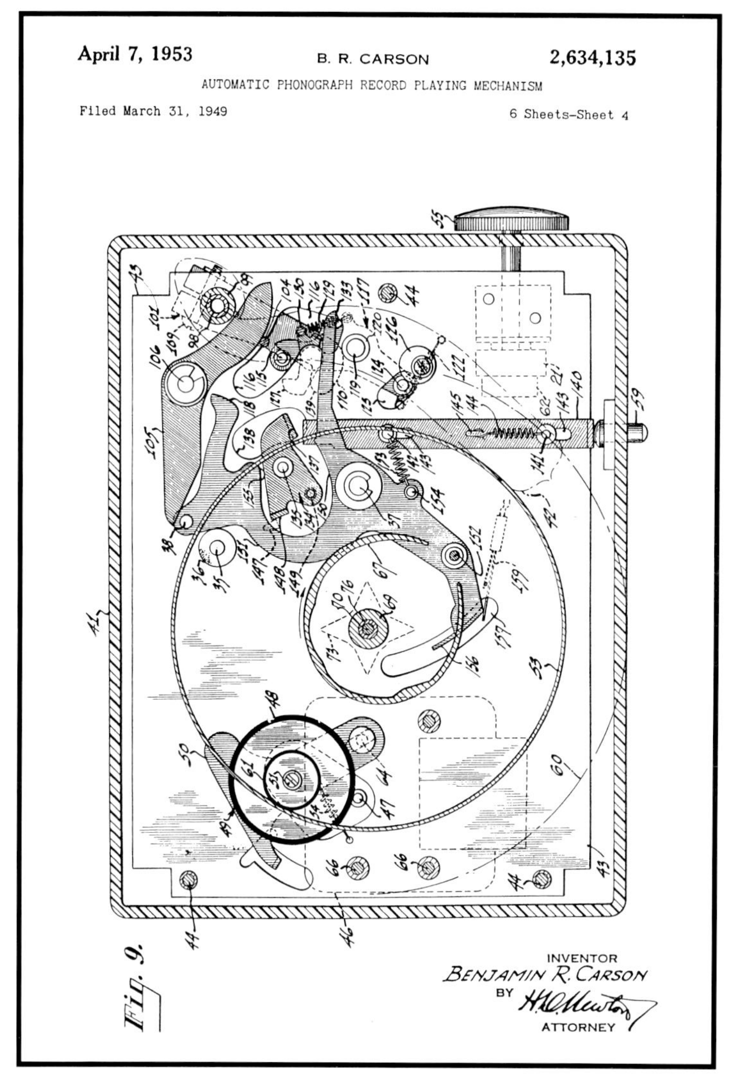

Carson's team finished the prototype and showed it to Joyce and Home Instrument's marketing department in March 1944. The group saw immediate advantages in the size of the player and the records beyond those specified five years before. The small size and reduction in parts meant the system combined with a radio could sell at $29.95, or half the price of the equivalent pre-war model. Size also enabled more choices in cabinet styling and installation in consoles. Joyce's group estimated that within a year of production, RCA could manufacture 925,000 changers in various formats, or more than six times the number Victor sold in 1940.[27]

Several issues remained to be negotiated with other parties before this could take place. First, "people would not buy the instrument if they thought they were going to be restricted by the manufacturer . . . to his own recordings." Therefore, "it would be wise to interest Instrument manufacturers as well as Record manufacturers . . . to make it an Industry development rather than a Company development." Victor's Home Instruments also had to gain Victor Records' cooperation in giving the new disc "a quality atmosphere." Already there existed cheap children's records six to eight inches in diameter pressed in a number of colored plastics. The X record was "small for other reasons. If possible it should look like a more costly product than it is."[28]

Finally, the Instrument Department's engineers opined "that the quality of reproduction . . . is commercially satisfactory." This was true for pop music and "all but a comparatively small part of the rest of the catalog." The engineers added that the process of developing the new record indicated the likelihood of their innovating "a 10" and/or 12" record which would have better quality than anything heretofore available commercially."[29] For the marketers, this meant reserving the possibility "at some later date, for an instrument designed for the possible High Fidelity Record—with reciprocal provision for playing 'X' type records." In the meantime, the prospect of a postwar rollout, when the industry's phonograph inventory was "near the vanishing point—the industry anxious for something new—the public anticipating something radical—seems particularly opportune."[30]

The Trouble with Changing Systems, 1942-1945

By the war's end in August 1945, however, the situation had changed. In truth, the assumptions behind the development of the 45 had been crumbling since Carson's proposal. Changes in corporate staff and structure, in competitive technologies, in the record industry, and in what the listener apparently preferred to hear all served to change the context in which Victor's prewar system would be introduced. Most importantly, a culture of high fidelity was emerging from disparate elements among consumers even as the company retained its assumptions about the reasons for innovation.

Strategy and Structure at RCA, 1942-1945

During World War II, the structure and management of RCA changed radically. Between 1942 and 1947, as RCA's management juggled military contracts and then postwar expansion, Carson's Advanced Development Group moved from Camden, New Jersey, to Indianapolis and back again. During the war the company stopped making all consumer products but one; at the Camden factory, "Everything but records was military. The plant was rearranged time and time again."[31]

In addition, the onset of military contracting resulted in the opening of the RCA Laboratories in Princeton, New Jersey, in 1942. Equidistant from the factories in Camden and Harrison, this independent research facility took the leading voice for high fidelity away from his allies in the Home Instruments and Broadcast Engineering Departments. Harry F. Olson, head of acoustics research, would continue to develop better loudspeakers after the war, but his isolation from the people working in a corporate profit center reduced his influence in maintaining RCA's reputation for sound quality.

With the relocations came organizational changes. At the end of 1943, RCA's board dissolved the RCA Manufacturing Company and Sarnoff hired Frank M. Folsom to run the RCA Victor division established in its place. Folsom, "a natural salesman," arrived from a career in department store marketing and wartime procurement with carte blanche to prepare the company for the postwar boom in consumer electronics.[32] In particular, if RCA was to reap the benefits of introducing television, its manufacturing branch needed better management than it received during the 1930s.

In one of numerous articles published in RCA Victor's inhouse magazine, Folsom likened RCA to "a large family"—if not a department store—which stayed strong "only if each member of that family is also strong, self-reliant and . . . self-sufficient."[33] To that end, after Folsom hired a Harvard Business School M.B.A. and management consultants to analyze RCA's administration, he decentralized RCA Victor in 1945, changing it from a functional to a product-based organization. Now, each product line, Records and Home Instruments among them, had separate engineering, production, and sales groups.[34]

This arrangement was necessary when anticipating a growth in the scale of production for the postwar market, but it marked part of the sea change in the operation and culture of RCA. Sarnoff's claim that RCA was a "company of men, not of charts," rang true in the 1930s. With no one directly responsible for the outcome of an innovation, RCA Victor's engineering, manufacturing, and marketing groups had to cooperate to develop new systems like television, facsimile, and the 45. With Folsom's change to a product orientation, developing new products became less important than selling the ones RCA already had.[35] Thus RCA Victor Records' advertising agency, J. Walter Thompson, which won the entire RCA account in 1945, did not find out about the 45 system until 1948.[36]

In addition, Folsom brought in his own people or new managers who had no allegiances to the RCA culture. One manager found that on returning to Camden in 1946, "people who had been superior to me in '43 were either dead or retired or absent, and there was a new group entirely, running things, who had never heard of me."[37] Frank B. Walker, who had succeeded Wallerstein in the Record Department, was promoted sideways in 1944 before MGM hired him to run its new record label.[38]

More importantly for the 45, Folsom's changes eliminated its support within management. Vice-president for advertising and promotion for RCA Manufacturing before Folsom arrived, Joyce was reduced to general manager in Home Instruments, reporting to Folsom. He could read the writing on the wall:

there would be no place in post-war marketing for the man responsible for pre-war efforts. As World War II ended, Joyce left the company for one of its distributors. Before he left, Joyce presumably tried to persuade his peer in the Record Department of the advantages of the 45 system. James W. Murray ignored it. Because of the restructuring, he had no incentive to coordinate marketing with the phonograph group; because of his affiliation with the new regime, he was uninterested in risking his position by backing his predecessor's innovation.[39]

In addition, Murray was no music lover. He maintained that the utility of records lay in hearing "any portion of a symphony or all of it," an attitude that reflected the company's increased emphasis on the middle-class, middle-brow consumer.[40] The importance of catering to the music lovers who once sustained the record industry declined as sales volume rose. The philistinism that accompanied this trend drove Charles O'Connell to resign as music director and increasingly redirected the music lovers toward Victor's new competition.[41]

The Rise of a Rival: Columbia Records and CBS, 1939-1943

Just as RCA Victor's management began mulling innovation in the winter of 1938-39, Wallerstein left to join the American Record Company. Now owned by William Paley's CBS radio network, American became Columbia Recording Corporation under Wallerstein in May 1939. Why he left is unclear.[42] Wallerstein played a crucial role in restoring the record as an entertainment medium and a vehicle for the diffusion of classical music in the 1930s. Restoring the Columbia label as a viable competitor to Victor offered a new and well-funded challenge. He brought with him the marketing techniques that made RCA Victor successful in the 1930s; RCA Victor's chief record engineer, James Hunter; two marketing associates; and perhaps an awareness of his former employer's intentions with the new phonograph system. Shortly after arriving at Columbia, Wallerstein began signing American symphonies and soloists as well as Benny Goodman and Count Basie. He also ordered that all classical recording dates be recorded in four-minute segments on transcription-quality sixteen-inch lacquer masters.[43]

Columbia's new head faced continued innovation in marketing from RCA Victor. Wallerstein's successor, Frank Walker, began sending out reviews of Victor records to newspapers and magazines; by 1941, seven hundred of the print media carried the free publicity.[44] That summer, Joyce arranged for the *New York Post* to offer coupons for anonymous classical recordings priced at fifty-nine cents apiece. As with the Victor Record Society, the ads included an offer for a cheap record player. As the campaign spread beyond the metropolis, Wallerstein and Columbia made a similar offer without the players in February 1940. Both labels acquired national sponsors to tie the campaigns to local schools and musical organizations.[45]

To keep Columbia from saturating the market with discounted records, Victor sued through the Publishers Service Company. Columbia withdrew the offer, but Wallerstein retaliated that summer. On August 9, Columbia announced cuts of up to fifty percent in the label's classical Masterworks records and packaged the discs in colorful albums. With a fraction of the business compared to RCA Victor, the company had little to lose in profit, and everything to gain in market share. In the next six months, Columbia's classical sales quintupled, helping double its overall sales.[46]

Victor had little choice but to match the price-cut. Joyce and Walker then cut prices on some Red Seal discs to fifty cents in the fall of 1941 in conjunction with the debut of Ben Carson's $425 "Magic Brain" console-changer.[47] While RCA pursued the high end of the market, Hunter introduced a record material that reduced surface noise by half.[48] This became a sales tool both in ads to consumers and to classical artists, who left Victor for a company where listeners could hear more of the sound of their performances. Wallerstein also expanded Columbia's pop music roster by hiring producer John Hammond to sign Benny Goodman and Count Basie.[49]

By 1943, Columbia's percentage of number one hits rose from zero two years before to thirty-six, while RCA Victor's share of the overall record sales had dropped by a third to forty-two percent.[50] The label's decline, the changes in RCA and the industry since Pearl Harbor, moved Victor management to hire its first full-service advertising agency, J. Walter Thompson, that year.

All This and World War, Too: Lower Fidelity and Higher Sales, 1942-45

RCA Victor lost its dominance as record sales rose, the American Federation of Musicians (AFM) went on strike, and the United States entered World War II. The strike and war forced the major companies to recycle older songs and older records. As a result, independent labels using non-union talent gained greater entry to the expanding market. The third leading company, Decca Records, signed a year before RCA Victor, signed several of its artists, and increased its market share. Since Victor was the last to settle with the AFM, the independents' supply of new talent and music helped squeeze Victor's control from below while Wallerstein's price cut squeezed Victor's margins at the high end.[51]

The industry's sound quality suffered from cuts in shellac and the wax used to record songs. The latter were made in Germany of a wax composition similar to that used by Edison in the 1890s. Victor stockpiled the material, but other labels began making recordings on aluminum and then glass discs coated with acetate-based lacquers. In Columbia's case, the switch proved beneficial to Wallerstein's 33 1/3 classical recordings, but Decca struggled to adapt old techniques to the new material.[52]

As British control of India came under pressure from Germany and Japan in 1941, most of the record industry was well prepared for the onset of rationing.[53] In April 1942 the War Production Board (WPB) ordered the industry to reduce the amount of shellac used by seventy percent, limited record production to 1940 levels, and fixed prices to those applied in December 1941. Until the WPB eased restrictions in April 1944, the industry had to make do with allotments equaling half of its prewar totals. To extend the supply of fresh shellac, companies lobbied against further cuts, limited the repertoire they offered, held record scrap drives, and encouraged dealers to establish trade-in programs for old records.[54]

The industry also sought alternative materials. One was a thermoplastic resin derived from phenol and decomposed bagasse, or sugar cane waste. RCA Victor used it to help recharge the thermoplasticity of shellac from scrapped discs.[55] This and other replacements like zein, a corn protein, and Vinsol, from pine tree stumps, were cheaper than vinyl and acetate plastics, which were unavailable by 1944 except for military purposes. The scrap and rising demand further diluted the use of fresh shellac and degraded the quality of the discs and their sound. Other factors, like the freeze on radio and phonograph manufacture in March 1942 and the diversion of skilled factory workers to war-related products, added to the degradation of sound quality in phonograph systems.[56]

When the WPB asked how much better records would be if the major labels received more shellac, the chief engineers from Victor, Columbia, Decca, and Capitol found it impossible to say. Until then, the record industry had little incentive for more scientific methods. Variations in scrap quality, inferior staff and equipment, and local practices prevented engineers from quantifying anything about 78s aside from their basic dimensions. One foreman determined strength by pressing a disc with his finger until it broke. Companies evaluated surface noise by listening tests since no one had found a way to measure it in relation to the type of music in the grooves.[57]

While record quality at home suffered for the duration, soldiers heard some of the future of disc records. To entertain the troops, the army's Special Services division commissioned RCA Victor to record and press "V-Disks." Victor made these twelve-inch 78s of pop, country, r'n'b, and classical music out of pure Vinylite. Without any filler, they generated minimal needle scratch; the limited press runs and care given to military products assured listeners of records with no manufacturing defects. The program began with a hundred thousand V-Disks in September 1943, and by war's end Victor and other companies were shipping over 330,000 every month.[58]

Deferring Innovation, 1944-1945

The changes in RCA Victor's management and the prospect of rival technologies for the consumer's ears and eyes contributed to the postponement of the 45. Despite the emphasis on television, Folsom wanted RCA to cover the home entertainment market in its entirety. Only fifteen to eighteen percent of radio owners owned turntables, and a significant part of RCA Victor's $40,000,000 postwar sales goal rested on increasing that percentage. That goal did not rest on innovation. Four months after Victor's marketers pondered the future of Carson's system in March 1944, Ben Aldridge of that group prepared two memos on the postwar phonograph market. In them, Aldridge evaluated the industry trends and the likely competition for phonograph combinations in light of the company's postwar planning. Despite his prediction that eighty-three percent of postwar phonographs would use automatic changers, Aldridge never mentioned the 45 as a counter to the competition's scaling up of production.[59]

The absence of the 45 had something to do with the new master of Victor Records. Aldridge was confident that RCA Victor could put over the system, but only with concerted effort, corporate support, and alliance building. Murray, however, was content with the status quo. Records offered music in "the most convenient and economical way . . . . in such a simple form that a child can make full use of them."[60]

Perhaps Murray based his opinion on the Commercial Research Department's survey of 4,300 consumers in mid-1942 about their use of records in anticipation of filling their postwar needs. In any case, Murray accepted the "perfection of automatic record-changing mechanisms of low-cost within recent years."[61] While Carson devoted his efforts to the changer of the future and to the changer for the elite market between 1939 and Pearl Harbor, other companies' engineers continued to seek solutions to the problem faced by the mass consumer seeking a changer a tenth the price of the Magic Brain. The problems Carson promised to solve with a new record in 1939 had already been solved, by other companies perhaps, but that was not Murray's concern.

By November 1945, Joyce had left and with him went the 45's last managerial sponsor. Even if he had stayed, the "'X-Changer' situation is considerably more complicated than it was the last time." Aldridge saw three obstacles in marketing the 45. First, the changing mechanism was not going to be as cheap "as had been expected." Second, sales of televisions and conventional combinations threatened to deter consumer purchases of a new record system for at least another year. Third, Aldridge feared a repetition of the format wars between cylinder and disc. Victor won that contest, but it might lose this one against the new media if "several . . . manufacturers decide to plug 'wire' and one, or several, decide to plug 'film' and if we alone support the 'X' Model—particularly if we do not have a strong, sustained, policy."[62]

The Trouble with Records: Other Needs, Other Answers, 1939-1948

The technologies Aldridge referred to received a boost during the war, as did others that offered better sound, more continuity, or both for the discerning listener. At the same time, it enabled improvements in the traditional medium for reproducing music. Equally importantly, the war raised the consciousness of a whole generation of men to the possibility of high fidelity, and gave them the skills and equipment to develop it. Finally, the war—and Petrillo's strike—sharpened interest in and demand for new classical recordings among music lovers, who took an increasing interest in the sound possible from records. These consumers, along with incipient audiophiles and veteran music lovers, organized sufficiently in the late 1940s to gain the attention of the mass media and the major record labels. Their success in installing high fidelity as a symbol of postwar modernity, where it had barely existed before, redefined the meaning of innovation in the record industry.

Two other technologies for reproducing music emerged from the war considerably stronger technically, commercially, and culturally. Frequency modulated (FM) radio and magnetic recording both benefited from military research and development as well as rising pre-war publicity. As the war wound down, station operators revived or established FM stations, while manufacturers promoted the benefits of new formats of older media that promised higher mark-ups and profits.

FM Radio

The other technology seemed to act as a rival suitor for the ears of music lovers and audiophiles. Along with television, as World War II ended, frequency modulated (FM) radio was being refined for the mass market. Credited to E. H. Armstrong in 1933, FM had inherent advantages over amplitude modulated (AM) radio in noise reduction, while its bandwidth allocation in the electromagnetic spectrum freed its transmissions of static and provided sound reproduction over 15,000 cps.[63] Armstrong promoted its noise-free, wide-range transmission of classical music to anyone who would listen.

He found supporters, but not among RCA Victor executives who, after the disappointment of "Higher Fidelity," saw television as much more salable. By the end of the 1930s, Armstrong had persuaded a number of entrepreneurs to establish the "Yankee network" of FM stations in New England and licensed his circuit designs to the General Electric Company. Despite difficulties in innovation of home receivers that rivaled those of RCA's television, FM had enough corporate, technical, and consumer support to call attention to the lessons of history. Either record executives would respond to the improvement in radio sound or they ran the risk of repeating the error of Eldridge Johnson in the 1920s.

Whether or not this was the lesson to draw from the relationship of radio sound and records, RCA's management responded to it over time in a variety of ways. One appeared to be Perkins's listener preference survey: would the mass market buy into wider frequency sound? Another was further investment in developing RCA's patent position in FM broadcast technology. RCA Victor announced plans to market the system after the war early in 1944.[64] By then RCA's George Beers had invented a circuit that reduced significantly the cost and effectiveness of FM receivers, and Stuart W. Seeley added another one in 1945. As the corporation's symbol of technological progress, RCA Laboratories endorsed the system because the expanded frequency range represented "high fidelity."[65]

Third was the application of FM to phonograph systems themselves. It was "essential that the . . . sound from records be at least comparable to or preferably better than that . . . from radio" if the record industry wanted to sustain its success. Charles Sinnett and other engineers developed and promoted a system using an FM-based pickup and circuitry at four engineering meetings in the middle of 1942. The system provided a frequency range of ten to twelve thousand hz "with an astonishing freedom from surface noise, mechanical noise, and distortion." The pickup weighed eighteen grams compared to the standard seventy grams of a conventional pickup. Finally, unlike the Rochelle salt crystals used in most electric phonographs, it was not subject to damage by heat, humidity, or changes in line voltage.[66]

Other members of RCA were less enthused about the promise of FM as a radio format. The corporation's relationship with Armstrong degenerated into a protracted legal battle over the inventor's basic patents. Ever sensitive to antitrust accusations, David Sarnoff and RCA had to balance the company's standing as a government-founded patent pool; the company's reputation for exploiting lone inventors; and the company's image as the pioneer of innovative systems using the electronic art, as Sarnoff called it. In this case, Sarnoff and RCA's broadcast network, the National Broadcasting Company (NBC), saw pure FM radio as less promising financially than either home facsimile or television. FM competed with these systems for spectrum allocation and the capital budgets of broadcasters as well as consumers. As with the promise of "higher fidelity" records in the 1930s, the acclaim for the wide frequency range of the system did not translate into a mass market. FM stations would never gain "large audiences" unless they "offered programs attractive to listeners."[67]

For the networks and most radio station owners, this meant broadcasting their AM content on their FM outlets. Some individuals acquired FM licenses in larger cities to play classical or instrumental background music, but few listeners found a more expensive receiver worth what one could already hear on AM. Other consumers bought tuners that turned out to "too cheap to be good," whose performance did not measure up to the promise. In addition, when the FCC changed the spectrum allocation for FM in 1945, 350,000 to 500,000 listeners with pre-war receivers had to buy new equipment.[68] FM radio remained a marginal medium for sound reproduction, but the existence of stations and small networks provided a node around which music lovers and audiophiles could gather.

Magnetic Recording

Magnetic tape or wire appeared to solve two problems that had plagued the phonograph since its invention. This was the technology that accomplished two tasks with equal facility—it could record and reproduce with no significant degeneration of the medium. Furthermore, wire or tape appeared to offer a more durable version of the waxed paper on which Edison had first tested the concept of phonography. They appeared to be an opening to continuous play of any length. Magnetic recording had been widely discussed before World War II, when the economics, if not the quality of the technology, appeared to indicate that magnetic recording and frequency-modulated radio were the media of the future.[69]

The military's need for durable recording media with those qualities stimulated development of magnetic recording and reproduction of sound. One researcher at RCA Laboratories worked on magnetic disc recording. The Armour Research Foundation provided a wire recorder capable of holding five hours of recordings on a spool of wire. General Electric sold portable wire recorders that were so popular with the military and reporters that it received over one hundred applications for postwar licenses. Other inventors demonstrated embossing techniques that offered hours of recording.[70]

After the war, however, manufacturers faced the transition from the military to the mass market. Wire and tape media were inconvenient to use because the consumer had to thread them through recording and reproducing heads and onto the take-up reel. Starting and stopping the reels always threatened to snap the wire, which frequently tangled. These structural drawbacks, together with the overestimation of the appeal of home recording, limited diffusion of magnetic media into the home.

Nonetheless, the Brush Company among other companies offered magnetic wire systems with what one observer described as "excellent fidelity." This assertion was based less on the frequency range—up to 5,000 cps at best—than the reduced distortion within that range compared to phonographs. These played from 15 to 180 minutes on reels costing from $2.50 to $8.25 an hour, which was comparable with the costs of 78s playing the same time. RCA Victor licensed Brush's technology and developed a thirty-minute wire cartridge played in a twenty-five pound machine.[71]

Several factors kept magnetic media from more than a tiny fraction of American homes until the 1960s. Beyond the complexity of these systems, most players weighed forty to fifty pounds and cost about ten times more than most turntables: $229 to $795. For Victor in particular, the potential of a format war with the 78 seemed to inhibit marketing. One editor observed that the recording industry, if it improved home recording to a point competitive with the phonograph record, would offer the unique instance of an industry that, "like the scorpion, stung itself with its own tail."[72]

Shaping the Status Quo: Record Consumption and Production, 1939-1947

Even as anticipation rose and faded over new formats, three factors contributed to the entrenchment of the 78-rpm record as the standard medium. Industry sales seemed to take even the industry by surprise. Manufacturers began to standardize the disc while developments in recording, materials, and playback equipment improved the quality of sound on disc. These changes helped deter RCA Victor's marketers from attempting to introduce the 45 even as they spurred the Record Division and RCA Laboratories to embark on methods of further reducing costs and improving the sound of 78s.

With the relaxation of controls on materials and production of records in 1944 and 1945, sales, prices, and profits began to take up where they left off before the war. Shellac prices shot from thirty-six to eighty cents per pound in 1946-47 before returning to pre-war prices.[73] Consumer sales jumped sixty-five percent in 1945 and doubled in 1946 before peaking at $224,000,000 in 1947.[74] RCA Victor's sales more than doubled in 1946, from $16,000,000 to $38,000,000, a figure matched the following year. At the end of 1946, RCA Victor turned a War Assets Authority factory in Canonsburg, Pennsylvania, into its fourth record plant.[75]

The increase had four consequences. First, much of the sales expansion came from new independent labels, whose share of the market grew from eight to twenty-eight percent between 1944 and 1947.[76] Second, it stimulated the rationalization of an industry still encased to a large degree by its nineteenth-century origins. Third, the explosion in sales also reduced RCA Victor's relative influence even as it moved the label to increase production and reduce costs. Finally, the success with the status quo reduced the incentive to innovate for the mass market.

In January 1946, as record sales continued to soar and, with them, the demand for record changers, the Radio Manufacturers Association offered a group of standards for consideration by its members. Number 169, "Proposed Dimensional Characteristics of Phonograph Records for Home Use," contained standards for thickness, diameter, stopping groove, and shape of the outer edge.[77] Given the variety of discs on the market and in consumers' collections, the standards did not resolve all of the issues surrounding 78s and changers. As time passed, however, it reduced the incentive to discard the 78 in favor of one company's innovation.

Shaping the Status Quo: 78 Research and Development, 1942-1948

On the other hand, the sound of the 78 became ever more fragmented because of the variety of recording standards and pickup technologies. The increase in new recording companies aggravated the variety in recording equalization. To limit overcutting of grooves of low frequency and high amplitude, recording engineers set a limit on amplitude up to a frequency between 250 and 800 cps. They also boosted higher frequencies that would otherwise fade or be lost in surface noise as the stylus approached the center of the record. In playback, an amplifier contained a circuit that compensated for the changing emphasis by boosting the low-end and compressing the highs, thereby eliminating much of the noise. Because record companies tailored their characteristic to what they regarded as the best phonograph, and because phonograph makers tried to build circuits that worked well with all records, there was considerable variation in the sound from a record, depending on which phonograph it played back. Even as engineers organized in the late 1940s to develop a single recording characteristic, by 1952, listeners contended with up to eight characteristics on their records.[78]

The playback industry also had to contend with dramatic changes in record pickups and styli. Since the advent of electrical reproduction, engineers had searched for ways to reduce the weight and wear of the steel needle. In the 1930s, Rochelle salt crystals made excellent transducers and reduced the weight of the pickup to two ounces. The reduced pressure encouraged the use of longer wearing osmium compounds for styli before World War II and then replaceable artificial sapphires afterward. The crystals suffered from humidity, however, and their output favored low frequencies. Continued research on electromagnetic transducers led to William S. Bachman's invention of the variable reluctance pickup in 1945. This device tracked the groove at twenty grams and reproduced frequencies out to 10,000 cps. At $6.95, GE's pickup helped build the market for high fidelity in tandem with new record materials and recording techniques. On the other hand, magnetically based pickups generated a thousand times less output voltage than the crystal format. Those listeners who desired the smoother and wider response had to invest in a new phonograph or modify their old one at considerable expense and with variable results.[79] After the war, then, record manufacturers faced a market owning phonographs, pickups, and styli of varying reproduction characteristics, pressures and wearability. The notion of high fidelity for the masses, had they desired it, must have seemed very distant indeed.

As for production efficiency, Victor had started this process in the late 1930s boom when its engineers began automating the pressing and timing processes.[80] Beginning in October 1946, RCA Victor dedicated the 6000 series of engineering memorandums to research in plating, stampers, record materials, production, and shipping. Despite the search for savings, the engineers never sacrificed the company's reputation for quality. Victor records were superior to their competitors, including Columbia's laminated discs, in resistance to breakage, warpage, and wear as well as in dimensional consistency and dynamic range.[81] Folsom echoed their commitment, for "quality and value" were "the cornerstones upon which brand name products firmly rest."[82] To sell the increased volume, Victor's market research group began testing self-service sales of records as early as 1942 in anticipation of "record supermarkets" after the war.[83]

The reputation for record quality extended to the material comprising the records themselves. The search for shellac substitutes during the war resulted in more experimentation and refinement afterward. The irregularity of shellac as a raw material, in price and composition, did not lend itself to the success of an automated mass-production industry. Engineers and chemists working under Hillel I. Reiskind developed combinations of plastics, extenders, and fillers that matched the qualities of shellac-based records in production and consumption. By 1948, they had completely replaced shellac with ethyl cellulose and Formvar vinyl acetate compounds at the Indianapolis plant.[84]

For Victor, competitive pressures at the pop end of the business served as an incentive to improve record quality. The movement into the business by Hollywood-based labels chipped away at Victor and Columbia's pop music market. The low barriers to entry, the competition for the next pop star, and the likely pressure on prices once the post-war boom eased "led to the widespread belief that to remain in business a manufacturer must produce a record of high quality [or] offer appealing innovations."[85] Thus Reiskind drew on the experience of producing V-Discs to produce a better-sounding 78. On August 30, 1945, RCA Victor announced its "De Luxe" Vinylite twelve-inch Red Seal record. Made of a proprietary blend of polyvinyl chloride/acetate copolymer and dyed a translucent red to identify the record itself with the trademark label, Serge Koussevitsky and the Boston Symphony Orchestra's rendition of "Till Eulenspiegel's Merry Pranks" retailed for $3.50.[86]

Some reviewers praised the discs for their low surface noise and "brilliancy" in sound.[87] Music lovers groused about the price, which was attributable to other factors beside the cost of the plastic. Vinyl compounds proved sensitive to the temperatures involved in pressing shellac records, resulting in more rejects in a press run. Second, the reduction in surface noise exposed flaws in the stampers; with tics and pops more apparent, engineers also rejected more stampers.[88] For these reasons, Victor released only three Vinylite albums in the first six months of the program. By 1947, however, as the factory crews mastered the production skills, the price dropped to two dollars a disc and the Vinylite record business grossed RCA Victor a million dollars a year.[89]

Despite the transition from natural to manmade plastics, engineers had not solved the primary complaint about sound quality for the mass consumer. Records continued to contain fillers that had generated needle "scratch" since the turn of the century. The use of mineral fillers continued for several reasons. Without the filler, the records would wear out much more quickly. Many consumers still used phonographs from the 1930s or earlier whose stylus pressure was measured in ounces rather than grams.[90] In production, fillers were far cheaper than manmade or natural plastics, and the postwar inflation encouraged their retention.[91] By 1948, finer fillers were available at no significant increase in cost, but they proved difficult to integrate with the plastics and with contemporary production processes. Moreover, the six- to nine-decibel improvement in surface noise was barely detectable and did not justify further research.[92]

Shaping the Sound of Music: The Emergence of High Fidelity, 1939-1947

RCA Victor struggled to balance increased production and quality with the diverse needs of its market within two greater trends. One was the growth of production; the other was the development of a consumer interest in better sound. The expansion of "high fidelity" technology, an industry to produce it, and an audience to appreciate it took place in large part as a consequence of World War II. Government funding of research on electroacoustics, production of precision components, and training of technicians provided the basis for a postwar boom in the field of sound recording and reproduction.

Defining high fidelity was and is a tricky proposition. As an experience, it represented an interface between the technical, the sensory, and the artistic. Since the 1920s, scientists and engineers had been tracking down the distortions and noises added to a sound signal in its recording and reproduction. When that signal was musical in nature and evaluated by the ears of a lay audience, however, the distinctions between flaws in the performance, the reproduction, and the perception became exceedingly fuzzy.

Finally, not all of the potential sponsors of better sound were convinced that the mass market preferred high fidelity. Repeated listening tests, reflecting the interests of the sponsors, indicated that most people preferred limited frequency range reproduction. Corporations did not speak with one voice on the issue, however. As it reorganized and diversified, RCA left open opportunities for engineers to develop and promote high fidelity.

The interest in higher fidelity existed well before World War II. In the radio industry, producers with investments in AM broadcasting touted their commitment to better sound in the face of increasing competition from Armstrong's FM format and its superior sound qualities. CBS touted its preparations for high-fidelity AM broadcasting, while the National Association of Broadcasters and the Radio Manufacturers Association sponsored a radio program that demonstrated the extension of broadcast frequency range from 1922 to 1939.[93]

The competition for the listener's ear also came from the phonograph industry, to which Hugo Gernsback's *Radio-Craft*, "Radio's Greatest Magazine," began to pay more attention. Writers and advertisers encouraged radio servicemen to expand their business by modernizing old phonographs or selling new disc recorders.[94] The magazine's contributors broke new ground in discussions on sound distortion, although not all of this was complimentary. One writer argued that magnetic pickups were "one of the worst offenders" in generating harmonic distortion. Another was among the first to explain the differences between the effects of harmonic and intermodulation distortion on increased frequency-range reproduction.[95] By 1941, interest in sound reproduction led Gernsback to categorize articles on the subject under "Sound" and to provide a monthly advice column on installations and equipment.

The challenge lay in extending this interest beyond the interest in public address systems or home recording, neither of which necessarily offered better sound than contemporary phonographs.[96] Much of the problem resided in the cost of better sound. In November 1940, Harry Olson, head of RCA's acoustic research group, explained the technical and economic challenges of high fidelity radio receivers. There was an inescapable correlation between higher prices and higher fidelity on the one hand, and declining sales volume on the other. Consoles costing over one hundred dollars comprised eight percent of the market; RCA's top models represented "only a fraction of a percent of total sales." Higher fidelity was available only if it was "accepted by the public and sold in considerable quantities."[97]

High Fidelity from the Bottom Up: 1942-1945

During World War II, the prospects for that larger market increased considerably. As early as June 1942, writers enthused over postwar prospects for the technology. Hundreds of manufacturers developed the capacity for mass production of precision parts, which could be devoted to better sound reproduction. The editor of *Radio*, the magazine for "Radio, Sound, and Electronics," predicted "a huge market for new and improved radio and electronic equipment." Both the public and engineers had "a right to hope for something better from radio in the post-war world."[98]

Military demand also led to the mass production of men trained in the use of electronics for radar, sonar, range-finding, and FM communications systems. Philco alone trained 25,000 technicians at its school in Philadelphia. A significant number specialized in the relationship of electronics to sound; two Western Electric Company engineers taught a series of courses on sound engineering in Los Angeles.[99] While the education helped create a postwar audience for high fidelity, many of those responsible for postwar innovations benefited from military employment or contract work. Robert Furst learned about FM radio for military applications and applied that experience to consumer receivers in the late 1940s. David Hafler's exposure to electronics in the Coast Guard led to a career in high-quality audio transformers. Arthur A. Janszen's work at the Harvard Underwater Sound Laboratory led to a career in acoustics research and loudspeaker innovation. Frank H. McIntosh's company became involved with industrial audio amplifiers, which led to the development of an amplifier that became a benchmark component for high fidelity. A. M. Poniatoff began his magnetic recording company during the war and commercialized the German Magnetophon afterward for use by radio networks and record companies.[100]

The use of electronics and recorded entertainment overseas also raised the consciousness of American soldiers to the notion of high fidelity. When the electronics specialists were not on duty, they took up the challenge of making phonographs that reproduced more of the sound on the V-Disks. The military did not provide for a playback system equal to the output of the custom-made records. Since soldiers in many areas had no power supply, the armed forces' Special Services branch distributed springwound phonographs. This "plywood squawk-box," as a veteran recalled the machine, had the same heavy pickup of his parents' generation to amplify the sound. At the same time it wore out the Vinylite grooves. One sergeant's foraging of parts in New Guinea to electrify a phonograph for V-Disks led to a career in customized home audio in New York City.[101]

High Fidelity After the War: 1946-1947

World War II expanded not only the audience for better sound, but the techniques and materials by which producers could fill that demand. RCA Victor's Vinylite Red Seal records were one offshoot. Curiously, Canby had nothing to say in 1946 about the other innovation frequently regarded as one of the two triggers for the high fidelity movement in the United States.[102] That summer, the British company Decca Records launched an American subsidiary, London Gramophone Corporation. London offered a recording of Stravinsky's *Petrouchka* using "Full Frequency Range Recording," or FFrr, on discs made of a high percentage of high grade shellac available through the English market. London also put on sale an inexpensive record player designed to play back that range properly.[103]

The process was a direct consequence of the war, developed by British Decca's chief engineer, Arthur C. W. Haddy. Haddy had been tinkering with high fidelity recording techniques since 1935 when, in 1943, the Royal Air Force's Coastal Command requested records that would reproduce the difference in the sounds generated by German and British submarines. In Haddy's words, "That meant high fidelity and the chance to conduct all necessary experiments under the best possible conditions. We went all out on it in a way that probably would have taken years in peace time." Most of this work related to the recording process. Haddy and his team improved the cutting head and developed a lateral-feedback recorder and amplifier that provided a flat frequency response between 40 and 15,000 cps. A year later Decca began combining this equipment with Wallace Sabine's research on reverberation times and microphone techniques for commercial recordings before it formally debuted the process on discs containing a high percentage of shellac after the war.[104]

High Fidelity from the Top Down: 1946

At the same time that the London disc and phonograph appeared, Henry Luce's magazine for affluent businessmen, *Fortune*, published an article under the innocuous title, "Music for the Home." The author began with the thesis that the "American public is getting a poor deal today in . . . musical reproduction." Despite the promises of wartime innovations, nothing was likely to replace the radio-phonograph combination. What then did listeners hear, and what could they expect to hear in the future?[105]

The answers were, not much, and little more, if the relevant industries maintained their attitudes toward sound quality. The writer described the regulatory and economic limitations on radio and the weaknesses of amplifiers, loudspeakers, phonographs, pickups, and records. These combined to limit sound reproduction to a noisy 5,000 cps. Combinations available for that Christmas offered undistorted reproduction up to 8,000 cps, the writer's minimum for high fidelity.[106]

Was the improvement worth the $400-$1,400 for a new console? The reader was given the perspectives of the "golden ears" by audiophile Thomas R. Kennedy, Jr., of New York City, and the tin ears in the collective judgment of the consumer by the broadcast and consumer sound industries. Kennedy, a radio engineer, was a purist. He damped his turntable in six hundred pounds of sand and monitored the effect of humidity on his Rochelle salt crystal pickup with a hygrometer. Kennedy dismissed suggestions of fetishism with the comment that anyone who listened to his system would "go home and throw rocks at your set."[107]

Assembled against Kennedy were the industrial interests with enormous investments in AM radio and coaxial network cables, patents, and factories. Even the writer seemed defeated by the evidence of mass-market preferences for limited fidelity. He could only assert the "musical public" was ready "to appreciate fine music" on "instruments of better design" even if ninety-five percent of professional musicians were not.[108]

Defining High Fidelity: The Music Critic as Consumer, 1946-1947

But what could people hear and what did they want to hear from their phonographs? The two questions were interrelated in a market culture and neither admitted of a straight answer. "High fidelity" meant better reproduced sound, but neither music critics nor engineers could agree to what the reproduction should be compared, or who should be the final arbiter of the quality of reproduction. On the question of preferences rested the investment decisions of the broadcasters and major manufacturers. More fidelity cost more money, and before they upgraded their facilities, they wanted to know whether the public cared as much about the form as the content of reproduced sound.

The most influential record critic mulled the first issue with some regularity. Edwin Tatnall Canby was raised in an affluent family, educated at Harvard in music, and employed at Princeton University to teach the subject. From prep school on, he had access to expensive phonographs and components, and at Princeton he fell in with the math and physics faculty who traded an education in engineering for one in music. Unlike most critics, Canby had foreseen the "whole 'audio' development, hi-fi and all, and the spread of recorded music" since the Depression. "Anybody could have seen it coming. I did. . . . What astonishes me is that everybody else didn't fall in too."[109]

In 1945, Canby started reviewing records for *The Saturday Review of Literature*. Unlike earlier critics who catered to the music lover clique, Canby tried to legitimate the recording and its criticism in three ways. The venue associated classical record consumption with book purchases and the many motivations and subjects for reading. Second, "music lovers" and "record collectors" became "record buyers" in Canby's writing.[110] Those who listened to classical records were no longer fanatics or obscurantists, but part of the larger American culture. Finally, by placing records in the context of changes in the industry, Canby made the field less esoteric. Writing easily with a patronizing self-assurance, he crisscrossed between critiques of sound and music. His reviews stood out because he rated "Engineering" under the subheadings of balance, realism, and record surface, as well as the artistry of the performance.[111] Canby made it his mission to promote "complete understanding and cooperation between engineer and musician." He would explain the issues involved in making a record to "the record buying public, which so often is unwittingly guilty of unfair judgments against both engineers and musicians."[112]

Canby treated the issue of high fidelity on a regular basis. In January 1946, he defined it as the "faithful reproduction of music" and stated that radio and record listeners "haven't had it." FM and better records preferred "genuine high fidelity—electrical reproduction that is . . . a reasonable facsimile of the original." Noting the public's conditioning by bad reproduction, Canby called for "a better understanding of the esthetic factors involved in perfect high fidelity reproduction" by the public and by engineers.[113]

By the end of the year, Canby was enthused enough about the promise of FM and phonograph technology that he urged his readers to buy only radio-phonograph combinations that had an improved pickup and speaker that could accommodate a frequency range at least double that available before the war. The key, however, was FM. It meant "a fundamental change in the standards of fidelity you will require in your new equipment" and in the sound that one heard.[114]

Canby also began contributing a column in 1947 to *Audio Engineering*, renamed from *Radio* in May. John H. Potts, the editor, turned the magazine into a forum around which audiophiles—amateur and professional sound engineers—could congregate and develop standards for the burgeoning industry.[115] Supported by a doubling in sales from five to ten thousand in four months and advertising from hundreds of component manufacturers, Potts focused on the relationship between music and sound. Beginning in June, he sponsored a seven-part "series of articles on music theory, written especially for sound engineers," that covered the subject from musical instruments and room acoustics to stylus-groove relationships and the human ear. As the question arose of what listeners preferred to hear, limited or high fidelity, *Audio Engineering* became the primary forum for engineers to debate the issue, Potts siding with those arguing for improved reproduction.[116]

Canby's "Record Revue," a feature "unique in a technical magazine," was included because "so many of our readers are record fans after hours."[117] It appeared to be an opportunity to cross the bridge he started building at the *Review*, but Canby found himself stepping gingerly between the world's best sound engineers and the world's best music lovers. A year after he first expressed his opinion on high fidelity in the *Review*, Canby was reluctant to give a firm definition to his new audience. The new pickups and improved recordings provided "almost limitless" frequency response and so little noise in the groove "that the listener is scarcely aware of the actual pickup and record at all—the music holds the field to itself."[118] By the end of the year, he merely applauded "every improvement in fidelity" without defining it. Instead he concluded, "There can be little argument, among sound men, as to what constitutes 'better' in general terms."[119]

Writing in the *Review* in October, however, Canby committed himself to "its original common-sense meaning . . . *a relatively high degree of naturalness, of faithfulness to the imagined original sound*." For the music lovers he went a step further, breaking the concept into two parts. The "illusion of reality" depended on the capture of the "rightly calculated sound" by microphone placement and acoustical control. While Canby acknowledged that much more took place in the transmission from microphones onward, this distinction left him asserting that "even the oldest" European recordings of piano pieces offered "a type of genuine fidelity that has not been matched in spite of our vaunted 'high fidelity.'"[120]

Canby found it difficult explaining to engineers the difference between aesthetic and technical fidelity that he had declared. In the next *AE* column, he emphasized only that all reproduction was an illusion because it referred to an "imagined original": those who heard the reproduction would never hear the original performance.[121] It was a point with which the engineers could not argue.

When he returned to the second part of his definition in the next *Review*, Canby went further. His enthusiasm of 1946 for wide frequency range disappeared. Music lovers who bought the new records and pickups would find that the rest of their systems revealed the problem of distortion, "in the higher tones" which previously had been suppressed. Conscious of his readers' budgets, Canby could not urge them to start over, or adapt their systems as the engineers did. Rather, he concluded weakly, "Don't misjudge 'high fidelity' until you've actually heard it!"[122]

Defining High Fidelity: The Inventor and the Producer, 1944-1947

Those at RCA who had a vested interest in the AM radio and the promise of television regarded the issue similarly. O. B. Hanson, chief engineer of NBC since its inception in 1926, was not opposed to FM. He agreed that a wide, undistorted frequency range was desirable but questioned the emphasis on 15,000 cps frequency reproduction. In a report to the wartime Radio Technical Planning Board, Hanson highlighted three physical problems in sound reception. The higher a frequency, the more directional it became, and few loudspeakers were built to disperse the frequencies off-center. How many people in a household could sit directly in front of the speaker? The lower the volume of the radio, the greater the drop-off at the low and high ends of the frequency range. Who listened to radio at a volume where they could hear the high and low frequencies above background noise? Who, for that matter, had ears young enough to pick up high frequencies at all? Hanson urged that everyone "get down to earth in the matter of high fidelity" and concentrate on resolving the distortions and noises present in contemporary radios rather than strive for "a theoretically complete audio spectrum."[123]

Hanson's call for a limited high fidelity reached the engineering and enthusiast communities in 1944 through a reprint in *Radio* and gained support from other authorities.[124] More writers joined in the effort to explain the relationship between frequency and loudness as well as other distortions generated by playback systems. The ear's perception of the relative loudness of different frequencies depended on the intensity of the sound. A bass drum or a triangle sounded much quieter compared to mid-range sounds when reproduced at low volume. Since most listeners did not play music at the level heard in a concert hall, engineers added tone controls to boost the bass and, increasingly, the treble, for household reproduction.[125]

But how was an engineer to determine where frequency and intensity compensation should begin or end? How were these circuits to give an illusion of fidelity to the original performance? One engineer agreed that, rather than attempt the impossible of bringing the listener to the performance, the industry should do the opposite. Listening to a monaural source at lower volume with greater background noise "cannot possibly 'sound like the original.'" Others urged their peers to "make a practice of occasionally attending symphonies and listening to live dance bands, with a conscious effort toward improving his audio discrimination." Another called for greater interaction between acousticians and musicians with the aim of developing a common terminology for music as an art and science as well as criteria for "measurement of music and its effect on people."[126]

CBS's Tin Ears vs. RCA's Golden Ears, 1945-1947

Not every engineer involved with sound reproduction felt as much concern. Like the smaller and pop-oriented labels, CBS and its subsidiary, Columbia Records, had little financial stake in better sound.[127] Unlike RCA, CBS did not manufacture radios or phonographs; it only broadcast content and manufactured records. If it was true that most people preferred lower fidelity, then CBS felt less obligation to innovate a better sound in the content that it sold. Like RCA in 1939, it undertook a survey to find out what the audience preferred as the war ended.

While RCA never made the results of its 1939 survey public, CBS showed no reluctance six years later. Philip Eisenberg and Howard Chinn, a psychologist and an engineer at CBS, collaborated on a similar survey that raised basic questions about the nature of sound reproduction. Categorized by age, sex, musical training, education, and FM radio ownership, nearly five hundred listeners indicated their preferences for different frequency and volume ranges in recorded and live broadcasts. No group expressed a majority preference for wide-range frequency response, although FM listeners changed their minds when informed that they had heard a wide-range sample. Because of the live samples and everyone's experience with high fidelity in everyday life, the authors rejected the suggestion that people had been conditioned to accept limited response. They could only conclude that "a narrow tonal range is preferred because it sounds better."[128]

Many engineers thought this conclusion counterintuitive. At a discussion during the 1946 meeting of the Institute of Radio Engineers, critics argued that the system used must have suffered from various forms of distortion, that listeners were conditioned to prefer inferior broadcast sound, and that the CBS tests implied most people would prefer live performances with similar restrictions. Surely "something could be determined experimentally about absolute preferences, and that . . . would be very desirable for . . . determining . . . the aim of the engineer."[129]

Chinn, Eisenberg, and their supporters pointed out that they had tested only monaural sound, not the binaural effect experienced at a concert. Yet they did not make clear whether they felt stereophonic radio was commercially unrealistic or that listeners' preferences would vary with more channels. Instead they refused to accept the notion of any ideal sound unless someone could provide "experimental or theoretical evidence." Scientists should pursue perfection as a goal but they asked, "What is perfection?" Was it a new and better device for reproducing sound, or was it "reproducing sound *the way people like it*?"[130] They released another study with more controls that supported the latter goal as more legitimate.[131]

Harry Olson and Acoustic Research, 1945-1947

The Chinn-Eisenberg studies served to reinforce suspicions that corporate producers opposed offering better sound to consumers. Manufacturers added inferior FM circuits to match those for AM in their radios.[132] During the debate within the engineering community, however, both sides were aware of other corporate research on what people liked to hear.[133] Olson was especially annoyed by the CBS conclusions. These tests, like the earlier ones at RCA Victor, Bell Labs, and NBC, were subject to at least six forms of nonlinear distortion as well as other forms of noise.[134] Well before Chinn and Eisenberg published their findings, Olson proposed a preference test using a live orchestra. If people preferred a limited frequency range instead of the 20,000 cps that Olson understood to represent the outer limits of human hearing, then perhaps "the public would like an orchestra better if it were surrounded by a . . . filter which eliminated the high frequency sound."[135]

Olson did not receive funding until 1946. After a thousand listeners listened to his band at the Princeton Labs with and without a filter, he announced the results. Over two thirds of the audience preferred full-range reproduction. Olson had looked "at the fundamental problem—whether people do or do not have a definite choice between . . . natural music sound compared to just a portion of it. The answer is, they do."[136] He concluded that the problem with earlier tests lay not with the listeners' conditioning but with the quality of reproduction, and his 1947 article helped revive research into record systems distortion.[137]

Among professional engineers, Olson's tests did not resolve either the question of definition or preference. In November 1948, the Acoustical Society of America convened a panel at its annual conference on "What Constitutes High Fidelity Reproduction?" The panelists included J. P. Maxfield, who had initiated the discussion with the rationalized electrical recording; Semi J. Begun, the leading authority on magnetic techniques; W. B. Snow, one of the Bell engineers responsible for the early stereophonic systems; and Olson. The first three assumed high fidelity meant an undistorted, wide-range reproduction of the sound entering the microphone; all accepted that the cost of this goal for the mass consumer was uneconomical. Therefore, as Snow proposed, engineers should work with listeners to produce "a satisfactory *impression* of the sound."[138]

Olson disagreed. In what the reviewer called "the most informative and original paper of the symposium," he explained how the components in $300 consoles could be upgraded at minimal cost to reduce the many distortions that alienated listeners from high fidelity. For Olson, a bridge to the ideal could be built to the general public more rapidly than many of his peers thought.[139]

Olson's work gained acclaim and support from the audio-related press.[140] *Audio* gave another engineer's "deliberately challenging" paper on the "beneficial commercial effect" of high fidelity. In it, C. J. LeBel highlighted the problem of "listener fatigue" that occurred from high-frequency IMD. The distortion was quantifiable; the fatigue, which occurred somewhere in the brain, was not. Nonetheless, it existed, based on experiences in hearing aid research and motion picture theatres, where the audience had no distraction from the reproduction. LeBel argued that producers should cater to the half of the population not satisfied with limited reproduction, for "every time a listener yawns and turns off his set, his ears have won a victory."[141]

RCA supported Olson's campaign. He had laid the groundwork to maintain the company's technical reputation and profit margins if the general public responded to the call of the audiophiles. The connection is more clear when his interpretation of listener preferences is related to RCA Victor's signal event of 1947. In August the division underwrote a train full of music critics from Manhattan to the Berkshire Music Festival in Tanglewood, Massachusetts.[142] There, they and twelve thousand music lovers heard and saw the Boston Symphony match performances with the RCA Victor's "Festival Series" of radio-television-phonograph combinations.

There was nothing new about this comparison except in the refinement in masking the deficiencies of sound reproduction. Thomas Edison had hidden the monophonic nature of his records by recording soloists; Olson arranged to play his monophonic system in the festival's Music Shed. Because of its open-air design, "the perspective of the orchestra is lost" for most listeners, meaning that they heard the symphony itself without a stereo effect.[143] Olson arrayed twelve of his Duo Cone loudspeakers at the front of the stage at the point where microphones had recorded the last four minutes of Beethoven's "Egmont" overture earlier. Powered by three forty-watt amplifiers, the speakers matched the acoustics and the sound level of the orchestra.[144]

The key improvements in the new systems lay in the speaker, which reproduced sound from thirty to fifteen thousand cycles over a ninety-degree arc with little distortion; and a "threshold noise suppressor" that Olson held to be an improvement over H. H. Scott's much-touted circuit.[145] No one was completely fooled but even some of the critics "were not quite sure which was performing."[146] All in all, "even the most cynical critic had to admit he had been impressed" at the sound that could be had for $1,800 to $4,000 from RCA Victor's new Consumer Custom Products Department.[147]

The Consequences for RCA Victor's Records, 1945-1947

While Olson upheld the notion of high fidelity and improved loudspeakers, others at RCA Victor pursued the problems of distortion in records. In 1945, H. E. Roys began studying the intermodulation distortion (IMD) generated by two frequencies reproduced simultaneously in a groove. IMD, like harmonic distortion, arose from the variation between the path traced in the groove by the reproducing stylus and that cut by the recording stylus. Mathematical analysis and tests with a variety of record diameters, stylus radii, and groove speeds indicated that listeners could not discern IMD of less than ten percent. Higher percentages arose from two sources. In the plating process that generated stampers from the original recording, the polishing of the metallized grooves at each step abraded the grooves. The extended use of stampers also produced records with distorted grooves. While these were matters for the manufacturers to address, Roys also found that current stylus standards resulted in "objectionable tracing distortion." He recommended the use of smaller-radius playback styli when manufacturers could mass-produce them. Carson's team applied Roys's work to the sound of the 45, while Olson's group put fine-groove "New Style Records" on its agenda in the spring of 1948.[148]

What the Consumers Heard: the Promotion of Columbia, 1945-1948

The favorable attention that Olson's and Roys's work received professionally did not translate into greater respect for RCA among audiophiles. The Berkshire line offered fine sound, but not at a price that most in that group could afford. More important, however, was the difference in philosophies between the audiophiles and RCA. Part of the attraction of seeking out high fidelity was the independence engendered by tinkering with and assembling the components to suit one's own tastes. Each man put his own stamp on the reproduction, becoming an engineer and producer in his own home. Canby catered to this desire, criticizing the convenience of consoles and calling for standard cabinets that could "house any and all components." For RCA Victor and other large manufacturers, this approach threatened their profit margins and trademarks. As Olson pointed out, cabinetry accounted for "a great deal" of the set's cost. RCA would lose control of the quality of sound reproduction if it built cabinets in which others could install equipment.[149] Olson may have been trying to extend the appreciation of high fidelity, but the interests of the corporation and the predilections of the audiophiles, as well the prejudices of their champions, left an unbridgeable gap.

The same was true in records, despite the efforts of RCA Victor's engineers in that department. In large part that was due to Edward Canby and the other critics at *The Saturday Review*. First, experiences as an amateur engineer notwithstanding, Canby did not understand all that went into a good-sounding record. Second, for all of his interest in bridging the gaps between musician, engineer, and consumer, Canby, and the *Review's* "Recordings" section focused on the classical record market and ignored the broader interests and strategies of the major record labels, in particular those of Columbia and Victor. As in the late 1920s, Columbia under Wallerstein continued to build its reputation from the top down. This meant attacking Victor's dominance in the classical field, which Victor maintained after World War II, and catering to the music lovers Wallerstein had worked with in the 1930s.

Columbia's efforts complemented those at RCA Victor. Franklin M. Folsom, whom Sarnoff promoted to executive vice-president of RCA Victor in December 1945, pursued a marketing philosophy based on General Motors research and his experiences at Montgomery Ward. Under his leadership, RCA targeted the middle eighty percent of consumers, the broad American middle-class.[150]

Folsom was thinking of economic groups, but the category applied to taste as well. As that view trickled down to James Murray and Victor Records, however, it also meant cultural levels. J. Walter Thompson, which took over the Victor Records account in 1943, advocated reaching a broader market for records, especially the premium Red Seals, despite the reservations of label executives during World War II. Maitland Jones, head of the account group in 1945, lobbied successfully with RCA Victor operating vice-president John G. Wilson and Murray, now vice-president for Victor records, to emphasize popular music and popular classics in order to expand the market for records.[151]

Thompson's strategy did nothing to endear Victor to the music lovers who had supported the industry for twenty years. The label continued to issue recordings receiving favorable reviews. But it focused less on the esoteric compositions sought by music lovers and more on the standard repertoire heard, usually as background to other activities, by the middle-brow audience.[152]

Canby's and J. Walter Thompson's effort to expand the classical records field became a battle of culture and commerce. Canby and the coterie that joined him at the *Review* struggled to imbue the newly middle class after World War II with the aesthetic and tradition of music appreciation that they had learned and developed in the 1930s.[153] The agency and its client sought to sell higher-priced records to the same group by promoting the most accessible recordings. As a result, Canby, Irving Kolodin, and others favored Columbia and its strategy in myriad ways. Canby concluded that Victor's shellac pressings had surfaces as fine as their plastic ones, even as he opined that the abrasives in the former were "on the way out."[154] Ignorant of the stamper's role in contributing to surface noise, Canby blamed the plastic for the "tiny ticks and sputters" he heard, finding them at least as annoying as needle scratch.[155] He implied that Victor deprived small labels of high-quality plastic, when it seems likely that the labels could not afford Victor's quality controls.[156] It never occurred to Canby that Victor's expansion of productive capacity made possible much of the smaller labels' output.

After Victor engineers took him to task for attributing a "vaguely disappointing" RCA Victor album to poor microphone location, Canby became more cautious in placing blame.[157] He continued to favor Columbia's records for a number of reasons, including its "exceptionally smooth-working team of musicians and engineers."[158] He praised the label's continued use of shellac "for sound reasons."[159] Despite recognizing that plastic was "a decided improvement . . . both in fidelity and in background hiss," Canby chastised Victor for not warning consumers that older pickups would wear the material faster.[160] Columbia's received more A's for its laminated surfaces than Victor did with its plastic or shellac discs in Canby's monthly evaluations. Shellac, he opined, was "probably more enduring than are any of the flexible plastics to date" and as late as March 1948 he found Victor's shellac surfaces superior to their Vinylite ones.[161] When Columbia began switching to plastic that spring, however, Canby wrote that its recordings, "with their wider range, justify the use of plastic."[162] Behind all of these technical complaints, Canby reflected the frustrations of other music lovers in the *Review's* Correspondence columns because Victor, unlike Columbia, held back on restoring its pre-war classical catalog.

For all of these criticisms, however, Canby and other critics found much to praise in the sound of 78s by the time Columbia debuted its LP. By late 1947 someone told him about the equalization techniques engineers had been using since the beginning of electrical recording. Recorded sound did not depend solely on microphone placement, which itself delivered "a new kind of sound, artificial perhaps, but very much alive and realistic nevertheless." Other factors existed that gave "an impression of decidedly greater tonal range."[163] Even in 1949, fourteen of the thirty-one orchestral and concerted recordings in the *Review's* annual summary consisted of 78 albums.[164]

Canby also predicted the future path of innovation in the record industry. In 1946, he inveighed against the "psychological jolts" provided by record changers and argued for a long playing record. The changer was "a deadly mechanism, and expensive." It destroyed not only records—one that broke Victor discs came to mind—but the peace of mind of the listener. He was forced "to listen to . . . the deadly rhythmical series of clicks and bumps that attend the changer's best efforts . . . tense and ready to spring at the slightest deviation . . . still tense with the possibility that the machine, after all, has merely begun the last record again, or even skipped a side."[165]

At that point Canby concluded that listeners should return to changing their records by hand if "we must (and we must) have 12-inch records, playing at 78 r.p.m."[166] After readers protested, he evaluated alternatives. Film, tape, and wire all promised "unlimited continuity" but were "least likely to replace our records" because replacement involved a systems change that neither producers nor consumers desired. Canby proposed something less radical: a revival of RCA Victor's 1931 radio transcription disc. After fifteen years of experience with the technology, he thought that the phonograph industry should be able to make a sixteen-inch long playing record and "a heavy, well-balanced, two-speed" turntable which, "with reasonable mass production, . . . should cost no more . . . than the price of a good changer."[167]

The only company making records and players, of course, was Victor, whom Canby never mentioned by name. As he saw it, the problem was that a convenience for the consumers "would involve a billion dollars worth of vested interest" by the company, "if not in actual manufacturing equipment, then in well established publicity routines." Until Victor or someone else resolved that difference in interests, listeners would "keep right on trying to patch [their] symphonies together from the familiar four-minute slices."[168]

CBS Invents the LP, 1944-1948

While RCA sat on the 45 and mulled improvements to the 78, CBS began working on the answer to Canby's dreams in 1945.[169] Columbia's announcement three years later came as a surprise to Canby and the other critics, in part because neither the impetus nor the funding came from Columbia Records. CBS's research director, Peter Goldmark, had been unhappy about the breaks in recordings and the space his record collection occupied. In November, after drawing on papers and patents of Hunt and Pierce on stylus-groove relationships, he showed executives Paul Kesten and Frank Stanton a record with twenty-five minutes of Tchaikovsky's "Violin Concerto in D" on one side. Turning the physicists' calculations into reality proved difficult, but the amount of distortion was less important to the executives than the prospect of entire symphonic performances on a single record. Such an innovation was of a piece with the promotion of CBS as the "platinum" network, in contrast to the middle-class domination of NBC and its larger number of stations.[170] Kesten budgeted $100,000 for development of a 33 1/3-rpm long playing record and rights to Hunt and Pierce's stylus design. The classical record and high-fidelity system would appeal to both classical performers and the high-culture audience frustrated with RCA Victor's middle-brow marketing.[171]

Wallerstein at Columbia Records was less enthused. In 1933 he had pulled Victor's long playing vinyl disc off the market because of its flaws. Now he refused to take the blame for failed R&D and regularly insisted that the record play longer and better. Goldmark received some help when he hired technician René Snepvangers from NBC's Transcription Department in New York late in 1945. Earlier, however, Snepvangers had worked on stylus-groove research for the FM pickup and the 45 in Camden.[172] Aware of the problems of tracing distortion in Goldmark's goal, Snepvangers warned Goldmark against a long playing record. By the end of 1946, Goldmark had put only eight minutes on a disc that Wallerstein found acceptable and Hunt rejected Columbia's price for a license on his pickup design. Columbia's head demanded more from "Goldmark's Folly," and this required skills that Goldmark lacked.[173]

He returned to work on CBS's color television system while Stanton hired an experienced engineer. William S. Bachman came from General Electric, where he had invented the variable reluctance pickup.[174] By September 1947, Bachman's team more than doubled the playing time and developed a tonearm capable of tracing the finer grooves without wearing them too quickly. That winter, Bachman's group attained over twenty-two minutes on a side, enabling the disc to play over ninety percent of all classical performances on two sides. Wallerstein insisted that he be able to hear the scrape of rosin on a violin bow and listen to twenty-eight minutes on a side, the amount necessary to contain any major symphony on four sides. At the same time, aware of RCA's 45 from Snepvangers, Goldmark wanted Columbia to invest in development and production of a 33 1/3-rpm single.[175]

After $250,000, however, Paley put his foot down. Never enthusiastic about capital investment for the sake of manufacturing, he was also dubious about the appeal of the innovation to the small number of record buyers who loved classical music or high fidelity.[176] There might be niche markets, but the single would require new presses, expanded plant to accommodate the transition from 78s, and the confidence that Columbia could lead the pop music industry, where sound quality was of minimal importance, to a new format.

The problem now became those of production and marketing. Hunter adapted part of the Bridgeport factory to cleaner standards than those used to make 78s. Since Columbia did not make phonographs, it licensed Philco to produce a two-speed player. Convinced that Hunt and Pierce would not have the wherewithal to fight Columbia in court, Bachman and Snepvangers designed a pickup based on their patents.[177] As to marketing, Columbia tried to make the LP as similar to 78s as possible, in color and dimensions to ease the transition from one format to a new one.

Vinylite lent itself to coloring. Hunter could have matched the discs to the royal blue label inspired by the use of blue-colored shellac in the Depression. Instead, Columbia retained carbon black as the coloring agent for three reasons. Colored records generated a sense of gimmickry and never sold as well as their black equivalents. After fifty years, record buyers, especially music lovers, associated black with stability and quality. Second, with the higher frequencies recorded on and reproduced from the discs, the graphite particles helped lubricate the grooves. Black discs wore less quickly and played more quietly as a result. There was another practical reason to keeping all discs in a plant the same color. Using dyes meant dedicating a production line because of the cleaning involved before and after a colored plastic; the down-time increased the cost of the records.[178]

Columbia also maintained the same diameters in the LPs as in classical and popular 78s. One team segued together the four-minute high-fidelity recordings onto twelve-inch masters for classical LPs. Others transferred the eight sides usually available in popular album of ten-inch 78s to ten-inch popular LPs.[179]

Since Columbia did not control the classical market, Stanton arranged a meeting with David Sarnoff to gain support.[180] The two executives lunched together in April 1948. Afterwards, Stanton played an LP in his office through a Harry Olson-designed loudspeaker. Sarnoff became "visibly upset" at the playing time as well as the sound. "I can't believe little Columbia Graphophone invented this without my knowing it," Stanton recalled him exclaiming. Stanton offered to share the technology if CBS could take credit a day ahead of RCA. Sarnoff seemed "genuinely interested" and Stanton arranged a formal demonstration.[181]

Sarnoff and sixteen RCA executives and engineers attended, as did CBS's owner, William S. Paley. After Stanton explained the agreement with Philco, Sarnoff dismissed the subordination of his company. He said that RCA would adhere to Columbia's design if it released a longplaying record. In the meantime, he invited the CBS leaders to a demonstration of RCA's versatility. At the next meeting a week later at Rockefeller Center Sarnoff showed off an RCA longplaying record, a tape recorder, and the 45. He argued vigorously against the introduction of the LP. He took great pride in the changer, "the best in the world, and the quickest," and the "kind of quality that a 7-inch 45 RPM record can produce and does." He also pointed out that the 45 promised to be cheap to produce and distribute.[182]

If this was intended as a threat, Paley and Stanton did not give in. Wallerstein grew anxious over the lack of agreement with RCA and the commitment to Philco. Stanton, desirous of luring Philco's sponsorship from NBC, refused to reconsider the June announcement. There was also the effect of another AFM strike, which held up new recordings all through 1948. Record sales declined that year for the first time since 1937. Debuting longplaying records featuring high-fidelity performances from years before made sense commercially. CBS would gain publicity at RCA's expense, and if the product was as good as promised, then a monopoly would ease the transition from novelty to product. Paley agreed. To defer any longer might allow Sarnoff to pre-empt their innovation with RCA's system; announcing now pre-empted the pre-emption.[183]

The Battle of the Formats, 1949-1950

In retrospect, the conflict between the LP and the 45 seems farcical. Of course the 45 was designed for single, mainly popular, performances at home and in the jukebox. By the same token, the LP suited the classical performances for which it was designed as well as collections of shorter pieces of all kinds. Why would either company try to make one size fit all?

The answer extends beyond the engineer's desire for standards or the executive's search for profits, or the record buyer's irritation with the conflict in the first place. Certainly, these had an effect on the outcome. But for RCA, the company with the most at stake in that outcome, the concerns were manifold. Contrary to popular belief, Columbia's early transfers varied widely in sound quality. Doubtless its engineers would improve on that, after which the label would seize the high ground acoustically and culturally. Without staking its claim, RCA Victor threatened to lose not only the classical market but the opportunity to supplant the 78 popular disc with its records and patented changer. Pushing the 45 as a classical format as well not only assuaged Sarnoff's ego but bought the recording department time to remaster its 78-rpm masters and offer an LP superior to Columbia's.[184]

Standing By the 45: Summer 1948

Despite RCA's silence, all was not calm. Should it stand by the 45, improve the 78, or make records the same speed as Columbia's? Over the summer, engineers at Camden, Indianapolis, and the RCA Laboratories in Princeton, New Jersey, argued with Rockefeller Center executives over the proper response. Support for the new speed wavered. Herbert Belar of the Labs' Acoustic Research Group had been charged with record research in conjunction with the Victor division just before Columbia announced the LP. He argued for a microgroove vinyl 78 based on sound quality and compatibility with customers' current collections. Standing by the 78 would leave an impression of conservatism, however, even if one made of vinyl with a microgroove sounded better and played almost as long as the LP. On the other hand, switching to Columbia's speed and standardizing a seven-inch disc with their rival would ease consumer confusion and pave the way for the two-speed players familiar in the commercial sector. By the end of the summer, the "X" record had evolved from "Original" and "Engineering" speeds of 45-rpm to a "Commercial" one of 33 1/3-rpm.[185]

Either move harmed RCA's reputation for leading-edge innovation. In addition, engineers argued that switching to LP speed for the small disc would result in a record with more distortion than the 45.[186] The ideal response would be to innovate ten- and twelve-inch longplaying 45s as well.[187] More importantly for the Home Instruments Department, even if the company gave up the patents on the record in order to gain acceptance, the changer represented licensing income from the smaller number of turntable manufacturers. In what some saw as the last hurrah of the engineers' influence at RCA, just before Sarnoff turned over the presidency to Frank Folsom, the product groups prevailed, and RCA Victor proceeded with plans to roll out the system.[188]

"Madame X" was the code name, during research and development, for an entirely new system of recorded music . . . perfected by RCA.

The remarkable background of **"Madame X"**

Now the identity of "Madame X," the *unknown* in a long search for tone perfection, has been revealed. From this quest emerges a completely integrated record-playing *system* – records and automatic player – the first to be entirely free of distortion to the trained musical ear . . .

The research began 11 years ago at RCA Laboratories. First, basic factors were determined – minimum diameters, at different speeds, of the groove spiral in the record – beyond which distortion would occur; size of stylus to be used; desired length of playing time. From these came the mathematical answer to the record's *speed* – 45 turns a minute – and to the record's size, only 6⅞ inches in diameter.

With this speed and size, engineers could guarantee 5⅓ minutes of distortion-free performance, and the finest quality record in RCA Victor history!

The record itself is non-breakable vinyl plastic, wafer-thin. *Yet it plays as long as a conventional 12-inch record.* The new RCA Victor automatic record changer accommodates up to 10 of the new records – 1 hour and 40 minutes of playing time – and can be attached to almost any radio, phonograph, or television combination.

Not only records are free of surface noise and distortion – the record player eliminates faulty operation, noise, and cumbersome size. Records are changed quickly, quietly . . . RCA Victor will continue to supply 78 rpm instruments and records.

This far-reaching advance is one of hundreds which have grown from RCA research. Such leadership adds *value beyond price* to any product or service of RCA and RCA Victor.

"Madame X" is finally revealed.
Courtesy Camden County Historical Society.

Production of the 45 players began in the fall of 1948. Even then it is possible Sarnoff would have accepted two new record speeds, one for classical and one for singles, since RCA engineers and marketers had considered the prospect of longplaying specialty disc since World War II. The LP's sound had been criticized, however, and Paley had begun negotiating with NBC's radio star, Jack Benny. On top of efforts by Peter Goldmark to obstruct RCA's electronic television, this was the last straw. Sarnoff was "deeply affronted" by the successful overtures to Benny, and it was his pique over the betrayal of his RCA family that provoked the battle of the speeds. For a year, RCA Victor would market boxed albums of symphonies in competition with the longplaying records of the rest of the classical industry.[191]

To some degree they must have been aided by Sarnoff's growing rivalry with William S. Paley, which the business press emphasized in covering the conflict.[189] Born beyond the Russian Pale, trained in the Talmud, and forced to support his mother and siblings as a child, Sarnoff acted on a strong sense of patriarchy in all of his relationships. Paternalism defined his conception of familial, corporate, and industrial roles. If the group rallied around the leader who set the standards, all would benefit from ensuing economic success. From this perspective, junior members, in this case Paley and the Columbia Broadcasting System, should not confuse the public with competitive formats in television or records.[190]

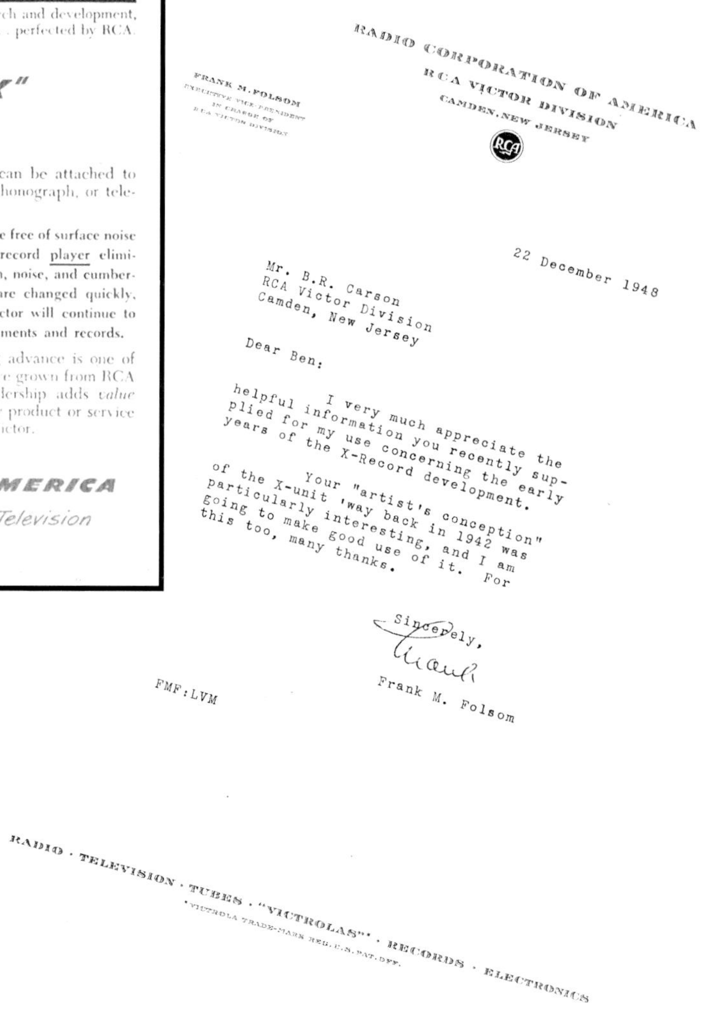

FRANK M. FOLSOM

RADIO CORPORATION OF AMERICA
RCA VICTOR DIVISION
CAMDEN, NEW JERSEY

22 December 1948

Mr. B.R. Carson
RCA Victor Division
Camden, New Jersey

Dear Ben:

I very much appreciate the helpful information you recently supplied for my use concerning the early years of the X-Record development.

Your "artist's conception" of the X-unit 'way back in 1942 was particularly interesting, and I am going to make good use of it. For this too, many thanks.

Sincerely,

Frank M. Folsom

FMF:LVM

RADIO · TELEVISION · TUBES · "VICTROLAS" · RECORDS · ELECTRONICS

Dec. 22, 1948 letter from
Frank Folsom to Benjamin Carson
regarding the X-record development.
Courtesy Camden County Historical Society.

Selling and Buying the LP: 1948

Columbia's announcement had mixed effects. The most objective record lovers who attended the press conference thought that Columbia's claims for "'full fidelity' . . . were not borne out. . . . Certainly this was not true when heard through a Philco machine where booming bass, low range and cabinet resonance existed." The use of an osmium point on the Philco stylus promised wear of the point and the groove after about sixty plays. The lack of breaks in two concertos played at home was "an interesting experience," but the 78s of the same performances "was more realistic and preferable to our ears." Even allowing for better playback equipment, the writer felt a year was necessary for judgment, especially since Columbia "has developed one of the best records on the domestic market."[192]

Another critic called the LP "the kid-glove record." Engineers expressed similar concerns. Like transcription discs, the LPs were likely to show "a substantial noise level if exposed to wear and dirt" and "should be noisier because of the lower recording level." Another warned of the need to maintain higher production standards because the margin for error was reduced with the finer grooves and slower speed.[193]

As for the market, manufacturers began inquiries almost immediately. Optimists at the National Association of Music Merchants convention held that week agreed that innovation had its best opportunity in a market off fifteen to thirty percent from 1947. Phonograph and record dealers expressed more reservations. Because the two groups did not often overlap, each wanted to know how the producer of the other component was going to make the system worth the dealer's and customer's investment. At the Home Furnishings Market in Chicago that followed the music convention in July, Philco showed off its two-speed phonographs and claimed that the six-gram pickup was the lightest ever made and offered the widest range of frequency reproduction. The "lack of comment" on the innovation was a "disappointing feature" to the market.[194]

Rumors of RCA's 45 helped defer dealer commitments. The Farnsworth Corporation, which had a cross-licensing agreement with RCA, announced plans to construct a 45-rpm changer, after which RCA, Columbia, Decca, and MGM all denied plans to produce records for that speed. Record dealers also resisted Columbia's efforts to claim that their franchises covered only shellac discs and to enforce quotas of the new records. Without players in people's homes, one dealer asked, "where are we going to dispose of all these LP's?"[195]

Two weeks later, Columbia's agency, McCann-Erickson, unveiled the campaign for the LP. Copy listed the advantages of the discs, including "a tone quality so lifelike you'll scarcely believe you're listening to a record!" The basic message, however, emphasized playing time: a full forty-five minutes. Consumers received "a complete symphony . . . on one record!" High fidelity was of less importance: one observer enthused that the LP was "the answer to the problem of the phonograph fan whose Bing Crosby collection alone spills all over the living room."[196]

For the audiophiles, however, Columbia gained coverage that touted the sound quality of the new records. *Audio* published information furnished by Goldmark in August and proclaimed that "engineers and music lovers alike are enthused" about the coming LPs.[197] An article in *Electronics* described the LPs as "transcription recordings for the home," aligning the discs with those used in broadcasting. The author glowed over the 10,000 cps frequency range, the 45 db dynamic range, and the low distortion. In all cases, LPs "outdistance shellac pressings," though Vinylite 78s might still remain superior in noise levels.[198] In *Radio-Electronics*, the writer came to an opposite conclusion based on listening to two LPs, that the "dynamic range is much wider than anything previously heard on records."[199]

Music lovers showed less enthusiasm for the sound even as they embraced the playing time. In the *Review*, C. G. Burke began a popular series comparing Columbia recordings in their LP and 78 formats. Burke's survey over the course of 1949 included tests for durability and price comparisons based on music per minute, but he focused on the sound.[200]

For dealers, promoting a new format proved a challenge. Aided by Columbia's props, stores put up window displays in which LPs seemed to come off best, despite the lopsided "tombstone" design used on the classical record jackets. LPs offered the basic appeal of "A whole symphony on ONE record!" and "uninterrupted performances."[201] The marketing had its effect. A dealer in New Haven, Connecticut, reported that he sold more LPs than 78s in classical music, which comprised sixty percent of its business.[202]

The mix of formats was of greater import to pop-oriented retailers. Most were upset over the confusion generated among prospective customers, the increased inventory, the price-cutting, and the need for more equipment to play back records in listening booths.[203] The forty-five minute playing time helped confuse customers who heard the rumors about RCA Victor's intentions. Among producers, Sydney Nathan, who turned King Records into the sixth largest label by 1949, had no interest in new speeds or vinyl. Nathan estimated that fifteen percent of the country's phonographs were still springwound. Many of their owners lacked electricity, and some of them preferred their machines' "horrible tone." This was his market for King's lower-class "race" and "hillbilly" records, and he saw no reason for change from the 78 as the pop music format "for years and years to come . . . further than I can anticipate."[204]

Responses to the 45: January-September, 1949

While Columbia seemed to move from strength to strength, RCA Victor struggled to organize a counterattack. The Christmas season would have been the best time to start sales, and Victor began remastering 78s and manufacturing 45-rpm players in the fall. Plans to make the records in Camden, however, fell through. The union refused to work with presses that stamped four 45s simultaneously in the time it took to press a 78. The machines, and RCA moved the operation to Indianapolis where Victor stamped the first 45, "Pee Wee the Piccolo Player," on yellow Vinylite, on December 7.[205] Later that month, RCA Victor began showing the system to the label's performers and the leading pop music labels, Decca and Capitol.[206]

As much as RCA Victor tried to keep the preparations secret, the meetings with Columbia gave its competitor an opportunity for leaks. Canby, writing in the *Review*, acted as Columbia's stalking horse. To offer a third system threatened "a suicidal chaos of competition" which would provoke "an enormous increment of ill will," implying that Victor already had something to atone for. Kolodin echoed Canby, adding the misinformation that the new record was not microgrooved. After scoring the unnamed label for its "Bourbonish purpose . . . to ignore the lessons of the past," he suggested that the goodwill derived from using 33 1/3-rpm would "much exceed the selfish pleasure that a battle of annihilation would give to a Pyrrhic victor."[207]

The apocalyptic tone signaled the fear of the music lover over RCA's size and the influence of its philistinism, accentuated by Columbia's self-portrait as the underdog. The music lovers assumed that with enough money, RCA could drive the LP from the marketplace, although the company had not driven another record format from the home since Edison abandoned the cylinder in 1929. The mass media joined in. *Newsweek* quoted the critics at length, and *Time* echoed dealers' fears, and took affront at Sarnoff's truckling to Petrillo in settling the AFM strike in December.[208]

Columbia's source of information about Victor's plans extended beyond the spring demonstrations. On the same day that RCA Victor demonstrated the 45 system to its distributors, Columbia announced it was producing a seven-inch Vinylite 33 1/3-rpm "LP" with a standard-sized spindle hole. Priced at sixty-three and ninety-five cents for popular and classical performances, respectively, these cost twenty percent less than the equivalent 78s. Columbia also announced that the Mercury Record Corporation would start pressing the larger LPs.[209]

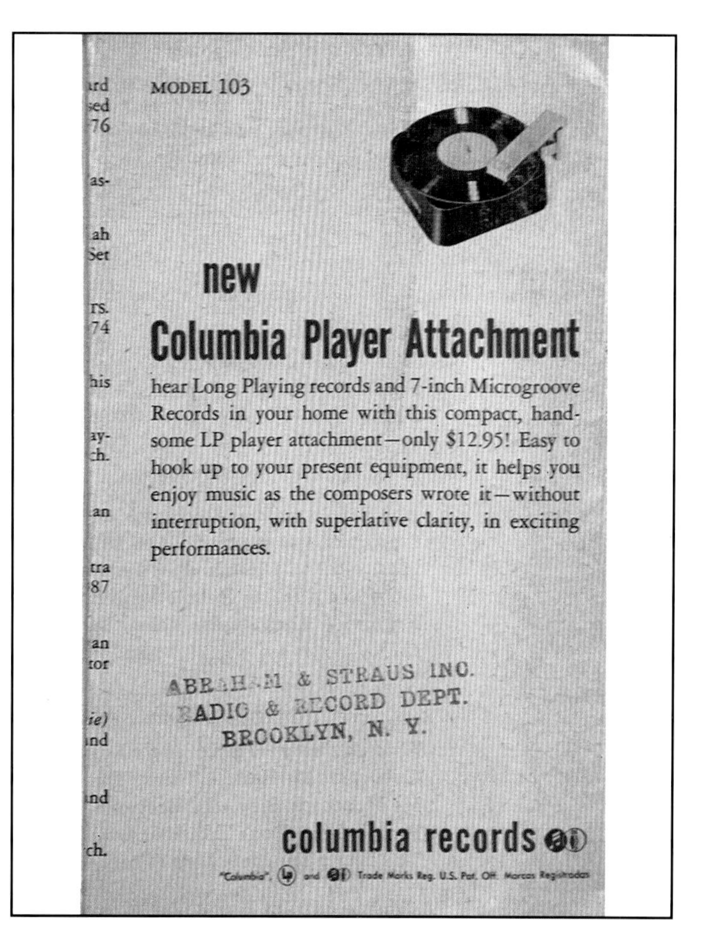

Booklet showing the new Columbia playing attachment. It could play LPs and seven-inch 33 rpm records, but only manually.

Columbia tries to counter 45 rpm introduction with a seven-inch pop 33 rpm record.

Pre-empted again by Columbia and ignored by the mass media, RCA Victor debuted the 45 for New York distributors and dealers on January 10, 1949, at the Johnny Victor Theater in the RCA Exhibition Hall in Manhattan. Folsom introduced the system as a "healthy ordinary competitive job that people do every day in business." One music lover disagreed with the benefit to the industry as well as the philosophy behind it, misspelling the RCA Victor president's name "Fulsom." Still, he had thought little of the LP's sound at Columbia's announcement and was impressed by Victor's staging. Engineers hooked up a 45 changer to a Berkshire phonograph to stake claim to clean reproduction up to 15,000 cps.[210]

The industry response to the 45 was not encouraging. Record and phonograph sales dropped forty percent in 1948 from the year before because of the AFM strike and the rising demand for television. Not all producers embraced the LP as a stimulus, but Columbia sold $5,000,000 worth of the format in four months, double what it expected. The music lover who praised the sound wished "more than ever" for standardization in the recording industry. High fidelity or not, the addition of a third speed, as one radio executive noted, was invitation to the chaos that the railroad industry would suffer from a change in track gauge.[211]

James Murray defended Victor's innovation in *The Saturday Review*'s "Recordings" section. In an ingenuous essay, he argued that RCA's unilateral effort at standardization served the public interest more than Columbia's. The company had tried longplaying records before and seen them fail in the marketplace. Without specifying what made the 45 the leading edge of the phonograph system, Murray asserted that the "distortion-free, noise-free performance" of the record "opens the way for . . . home instruments of wider frequency range and truer fidelity." Murray ignored his audience and emphasized Victor's interest in "satisfying the largest segment of the public which constitutes the . . . market for records."[212]

The music lovers rejected his logic. Victor's 1931 innovation had only playing speed in common with Columbia's LP. Editor Irving Kolodin countered that the choice of standards was one of diameter—the 45's 6 7/8 inches—versus playing speed. Why tie classical performances to Victor's procrustean bed of five minutes when Columbia's system allowed labels to use "as much of a record's surface as their needs require"? To support his reasoning, Kolodin printed six letters also rebutting Murray's explanation and noted that none of the many other correspondents supported RCA Victor's innovation. Writers to *The American Record Guide* expressed similar irritation.[213]

For records, Victor started with four singles, including the Boston Pops Orchestra's rendition of "Gaité Parisienne" and Arthur Crudup's "That's All Right." A series of singles and albums followed as recording engineers remastered 78s from the MALB catalog.[214] The term "album," which had lost its literal meaning with the debut of the LP, suffered further in the case of the 45. These were now stored inside a box thick enough to gave sufficient room to label the edge with half-inch print.

Somewhat to his surprise, Kolodin found himself congratulating Victor's engineers on the quality of their transfers. The 45s' surfaces were "uncommonly quiet," more so than LPs, if prone to warpage, and "the smoothness and response to the Red Seals are outstanding." Like the LPs, the sound correlated to the original recordings. Having played the discs on an adapted player, the "Recordings" editor reserved judgment on the changer and its effect on appreciating the twelve sides of Tchaikovsky's "Sleeping Beauty."[215]

Burke took up the issue four months later. He refused to compare 45s to the LP, except in the issue of continuity. Compared to Victor's 78s, the 45s excelled "in tonal quality" but the relative newness of the standard releases reduced the difference shown up between the LP and 78s. Otherwise "the Victor novelty consists of an obsolete principle (short duration) complicated by defective mechanical application." Carson's efforts to save on material left insufficient margin for production flaws, and Victor recommended masking tape on the labels to keep the discs from slipping when stacked under the pickup on the turntable.

Technically, the discs received high marks. An anonymous engineer in *Audio Engineering* began the review process with information from RCA licensees and other record companies.[216] "Stylus" described the 45 as "a more compact, lower-cost record of higher quality . . . . aimed chiefly at the popular market—which includes 80% of all disc sales." After praising the changer, he tackled the issue of sound quality when compared to Columbia's single. Other factors being equal, the 45's higher speed enabled a wider frequency range and much less tracing distortion as the needle approached the center.[217]

Benjamin Carson poses with the 45 rpm phonograph he designed after its introduction.
Courtesy Camden County Historical Society.

Benjamin Carson also designed the automatic 45 rpm record press. Having six heads, it makes a 45 rpm record every 2.5 seconds. It uses hydraulic water pressure at 1800 psi. Picture is dated 1943. *Courtesy Camden County Historical Society.*

Stylus refused to compare the 45 with the twelve-inch LP because of their different markets and prices. When pressed, he showed no love for music lovers. Classical records were "a mere excrescence, a wart on the surface of the record field. . . . Is there any reason why 80% of the market should take inferior engineering to please 20%?" Another article by the writer reinforced this point by showing how 33 1/3-rpm recording also contributed to degraded high-end response. When all was said and done, however, Stylus echoed the opinion of Belar and Olson's lab. Why not produce a longplaying 78? With a groove pitch similar to the LP and 45, companies could put nineteen minutes on one side of a twelve-inch 78, where the "quality of the sound would be superior to that on the LP at the same diameter."[218]

The battle among the inventors settled down to distortion. Just how much better were the new discs to their counterparts or the 78? Carson, Burt, and Reiskind highlighted the challenge of avoiding more than ten-percent intermodulation distortion. Roys had established that as the level noticeable by most listeners, and it became a sales point for the "live-talent tone" in ad copy. Al Pulley, Victor's senior recording engineer, gave a three-hour demonstration to the *Record Guide's* Peter Hugh Reed, who noted the improvement over 78s within the player's limited frequency range as well as the potential with a wide-range reproducer.[219] By comparison, Goldmark, Snepvangers, and Bachman withdrew their professional article on the LP when they learned of the specifications for the 45. Unable to dispute the problems of inner-groove distortion of any kind, they resorted to highlighting the superiority of the LP's sound compared to the worst levels of harmonic distortion produced on a 78.[220]

The arguments among the pure engineers missed the point, however. Canby was not alone in stating that, aside from the many other ills that beset the LP, Columbia's invention attained a higher aesthetic fidelity by reducing the number of unintended breaks in a performance. This improvement contributed to a higher technical fidelity. The LP experience took "recorded music right out of its own field, into the 'actual performance' feeling of a radio or a live concert." Nearly fifty years into the habit, listeners found that those "familiar disturbances . . . suddenly turn out to have been of vital psychological significance." Continuity overwhelmed ticks and pops, wow, the delicacy of the grooves, and the IMD on the inner grooves of the discs. In these conclusions, musically inclined audiophiles and music lovers alike agreed.[221]

Marketing the 45: March-September 1949

All of this added to the problems of marketing the 45. J. Walter Thompson's staff became increasingly frustrated over the budgetary and philosophical limitations to its role. They pointed to the fifty percent of consumers who did not own phonographs and the few record purchases of those who did as indicators of an untapped market. RCA Victor should help dealers by pushing a new audience to their stores. The ascendance of Folsom, who was fully confident of his own marketing expertise, and the persistence of traditional methods of retailing records aggravated conflicts with the agency.

The unexpected death in 1950 of vice-president John Williams, who supported Thompson's approach, contributed to the agency's loss of influence.[222]

While Victor's new president must have shared Thompson's belief in marketing to the consumer, Folsom had his own ideas about the method. He summed this up in the "Folsom Formula." All advertising promoted the trademark, and the quality and advantages of Victor's products. The emphasis on value reflected Folsom's Catholicism, his upbringing in the far west early in the century, and his sales experiences during the 1930s.[223] It is difficult to argue with Folsom's success, but his approach limited an agency's role in associating desire for a product with other experiences or encouraging consumption for its own sake.

Victor's Record and Home Instrument Department executives seemed to share neither Thompson's philosophy nor methods. They had tradition on their side. No more than fifty percent of American households had ever owned a phonograph. Radio surpassed phonographs in household diffusion during the Depression, and television would do so by 1953.[224] In the phonograph industry, distribution and sales traditions were not that far removed from the Victrola era. Dealers no longer sold talking machines door to door, but the devices remained an item for the young, bought primarily in the months before Christmas.[225] Victor's research and development notwithstanding, self-service record shopping had not yet come into its own. Consumers still bought most records through independent dealers who offered playing booths to sample the music and sound before purchase. The bulk, cost, and risk of records limited any one store's inventory.[226]

The 45 system went on sale in New York City on March 31, 1949, supported by the area's largest distributor, Bruno, which was owned by David Sarnoff's brother Irving. In a low-cost campaign, RCA then sent teams of salesmen to its other distribution centers across the country. RCA's control of its distributors and their supply of televisions forced dealers to add the new phonograph system to their stock. It was a cheap and easy alternative to Thompson's ambitions, for it passed on the problem of inventory even as it hurt dealer relations.[227]

Ad from May, 1949 issue of *Record Retailing*.
Courtesy Camden County Historical Society.

Yes, NEW record buyers!

Starting with June,
RCA VICTOR monthly releases
will be for both
standard 78 rpm
and
the new 45 rpm
'Red Seal'
recordings

YES! Thousands of *new* record player-owners, *new* record buyers are being created—people who are acting on the big RCA Victor 45 rpm advertising and publicity campaign now in full swing. So now—sell *both* these new 45 rpm buyers *and* your regular 78 rpm customers! Look at this line-up for June—give it a bigger-than-ever promotion!

Boston Pops Orchestra, Arthur Fiedler: Holiday For Strings; Our Waltz —Rose. 10-1311, $1. 49-0407 (45 rpm), 95¢.

Robert Merrill: Folk Songs of The British Isles (Phyllis Has Such Charming Graces; My Lovely Celia; Mary of Allendale; Down By The Sally Gardens; Come, Let's Be Merry; The Ballynure Ballad; Oliver Cromwell). MO-1306, $4.00. WMO-1306 (45 rpm), $3.35.

Yehudi Menuhin: Habanera (No. 2 From Spanish Dances, Op. 21)—Sarasate; Scherzo Tarantelle, Op. 16—Wieniawski. With Gerald Moore at the piano. 12-0922, $1.25. 49-0404 (45 rpm), 95¢.

Whittemore and Lowe: Coronation Scene (from "Boris Godounoff")—Moussorgsky; Polka (from the ballet, "The Age of Gold")—Shostakovich. 12-0923, $1.25. 49-0405 (45 rpm), 95¢.

Indianapolis Symphony Orchestra, Fabien Sevitzky: Ballet Music of Delibes (Ballet Suites: Coppelia and Sylvia). DM-1305, $7.25. WDM-1305, (45 rpm), $5.25.

Boston Symphony Orchestra, Serge Koussevitzky: Serenade No. 10, in B-Flat, K. 361—Mozart. DM-1303, $6.00 WDM-1303 (45 rpm), $4.30.

NBC Symphony Orchestra, Arturo Toscanini: Concerto No. 1, in B-Flat, K. 191—Mozart. DM-1304, $2.50. WDM-1304 (45 rpm), $2.20.

Aksel Schiøtz: Flow My Tears; Shall I Sue; Now Cease My Wandering Eyes—Dowland (From the Second Book of Ayres, 1600). 12-0924, $1.25. 49-0406 (45 rpm), 95¢.

All prices are suggested list, subject to change without notice, exclusive of local taxes. Prices of single records do not include Federal Excise tax. DM albums also available in manual sequence, $1 extra.

The world's greatest artists are on

RCA VICTOR RECORDS

Early illustration of the new model 9EY3.
Courtesy Camden County Historical Society.

Never before such clarity and brilliance in recorded music!

A COMPLETE PHONOGRAPH ONLY 10 INCHES SQUARE! There is also a compact plug-in unit, with which you can enjoy the new records through your present radio or phonograph or television combination. And there are other table and console models . . . all handsome, compact, and all with the famous RCA Victor "Golden Throat!"

How, then, was a dealer to lure prospects to hear and hear about the first new record since Victor's longplaying and picture discs in the Depression? Unlike the Orthophonic system's debut, the 45 had no player capable of demonstrating a clear advantage in sound over the best 78s. A rival innovation that offered a unique advantage already existed and enjoyed support by other companies.

The answer was novelty in appearance, to provoke curiosity and desire. Dealers' windows for 45s had no slogan. They promoted a system, reminded the consumer of the need for a new player, and featured rings of colored vinyl discs. The first 45s appeared in colors chosen by industrial designer John Vassos as "characteristic of seven classifications of music": black for popular, red for classical, yellow for children's, cerise for folk, green for country western, blue for "semi-classical," and light blue for "international." Victor touted the colors for adding "eye appeal to ear appeal in the playing of recorded music" and easing sorting by dealer and consumer.[228]

Black vinyl "Popular" 47-series. Later on, the blue label was changed to black.

Red vinyl "Classical" 49-series in a cellophane envelope (the first envelopes or sleeves for 45s). *Courtesy Jim Apthorpe Collection.*

Green vinyl "Country and Western" 48-series. *Courtesy Jim Apthorpe Collection.*

Cerise vinyl "Rhythm 'n Blues - Spirituals" 50-series. Rarest series of early 45s. *Courtesy Jim Apthorpe Collection.*

Yellow vinyl "Children's Little Nipper" circa 1950.
Courtesy Jim Apthorpe Collection.

Light blue vinyl "International" 51-series.
Courtesy Jim Apthorpe Collection.

Midnight blue vinyl "Semi Classical" 52-series.
Courtesy Jim Apthorpe Collection.

The campaign became a nightmare for Victor and its dealers over the summer. The records in the windows sometimes suffered from the sun's heat. Worse, the colors reinforced the impression of artifice and cheapness that plastic represented to most consumers. The colors also bogged down production by complicating the changeover from one record to another in the Indianapolis plant.[229] RCA engineers and advertising claimed that the changer, the basis of the innovation, was the fastest ever, although reviews suggested otherwise.[230] Pricing also ran against acceptance of the new speed. The changer and phonograph retailed for $24.95 and $39.95, respectively, or more than equivalent 78 or LP players. The records saved storage space but, at 95¢ for Red Seals and 65¢ for everything else, only the premium discs represented a significant discount from 78s.

As for the Red Seal marketing, various parties persuaded maestro Arturo Toscanini to pose for a series of photographs while he listened to one of his performances played on the "jewel-box" phonograph.[231] Victor never used those photos, however.

At the same time, Columbia pressed its advantage in providing "uninterrupted music at its finest."[232] In June, it began promoting a record player for $9.95. By July, Wallerstein reported that Americans had bought 750,000 LP players and three million LPs.[233] Dealers cut 78 prices in half, and estimated in one case that Columbia's LPs were outselling Victor's albums of 45s by a ratio of 30:1. While Victor promoted Vaughn Monroe's "Riders in the Sky," the best selling pop song in the spring of 1949, its failure to exploit the popularity of Broadway musicals haunted them. Columbia issued the soundtrack to *South Pacific* on LP at the same time, sales of which assured other labels that the format appealed beyond the classical market. By September, Decca and London, along with fifteen classical labels, had joined Mercury in pressing LPs of popular and classical content. Only Capitol joined RCA in making 45s while it also pressed LPs. When the convention of music merchants met in July, Columbia's sales manager gave a talk over the background of an LP side of Tchaikovsky's "Nutcracker" suite. The effect was impressive, and in a straw poll of speeds, dealers voted overwhelmingly for the LP system.[234]

Consumer Choices, 1949

For the consumer, the situation was more complicated. Should she—for Victor's advertising featured women and their hands holding the delicate discs almost exclusively—buy one system and stop listening to the other label's performers? Should he buy both and "clutter up his living room [and] his record cabinets," or stand pat and stick to 78s? In the fall of 1949, one consumer group's engineer evaluated the three speeds and five formats: shellac 78s, Vinylite 78s, 33 1/3 LPs, 33 1/3 singles, and 45s. He had little good to say about either new speed. The LP emitted "a peculiar 'spitting' type of distortion which is objectionable to the musical ear." Most pressings were poorly centered, resulting in "excessive 'wow'" while the "clicks and pops" generated by an LP's surface were "more disturbing than a low-level steady hiss" given off by a 78. The

45, "with economic factors duly considered . . . is a good idea" and provided less distorted reproduction. But the cleanliness of the recording made the surface noise and motor rumble more obvious. The cheapness of the phonograph and the limited frequency reproduction—70 to 4,000 cps—made Victor's choice of a new speed especially puzzling. When one considered all eleven factors involving cost, sound, repertoire, durability, standing pat appeared to be the best bet. The engineer suggested, however, that newcomers buy into the 45 if they were "not primarily interested in the highest grade of reproduction. . . . For serious music lovers" a 33 1/3-rpm record-playing attachment would take care of the occasional LP.[235]

By May 1949, when Columbia had been selling LPs for nearly a year and RCA Victor the 45 for six weeks, dealers outside New York City reported that the new records comprised only fifteen percent of sales.[236] The variety of formats left record sales floundering, but we should not ascribe the decline in sales entirely to indecision over record speeds. The AFM strike in 1948 and a national recession in 1949 held up sales, first in records, and then across the economy.[237] In the home entertainment sector, television set sales accelerated from one to three million in those two years, and would continue to siphon off consumers' disposable income. Finally, the major labels' control of the popular music industry itself fell in a slump between the end of the big bands and the rise of rock 'n' roll. From 1948 to 1952, Victor had only six of the forty-eight top-selling discs, compared to thirteen in the previous three years. Columbia and Victor both cut prices on 78s in 1949 and started or revived discount labels that summer. Record sales did not recover 1947 levels until 1955.[238] The irony was that records as a medium for music slumped just as popular interest in music swelled. Spurred by an industry push for music education, an amusements tax that reduced the number of dance halls after the war, and the increase in broadcast and recorded music, instrument sales and interest in music boomed in the late 1940s.[239]

Marketing the 45: September-December 1949

Sales of 45s went from bad to worse over the summer. Victor's share of the record market fell from twenty-one percent in 1948 to well under fifteen percent in 1949. RCA Victor and Columbia's executives met to find a compromise. Columbia held the upper hand: by September, Victor was the only major label not making LPs while only Capitol joined in pressing 45s. The ten- and twelve-inch records were used for both the prestigious classical performances as well as popular albums. In such a situation, agreeing with Columbia that both labels would press both 45 and 33 1/3 rpm records seemed the only way out.[240]

J. Walter Thompson and the Home Instruments Department disagreed. Since 45s could be adapted to small spindles, phonograph manufacturers would make only small-spindle changers, obviating the need for the 45 system and the 45. The company would lose its reputation completely unless RCA went all-out to convince the industry, dealers, and consumers that its system "was here to stay."[241]

The decision to follow Thompson's advice probably involved Sarnoff as well as Folsom. In the context of corporate pride, RCA's imminent involvement in government hearings over the future of color television, the patents on Carson's changer, and the inherent advantages of the 45 over popular 78s, RCA renewed its investment in its system. Over the twelve months from September, Thompson and RCA embarked on a marketing campaign that demonstrated the "effectiveness of hard selling in this type of situation."[242] Over the last three months of the year, dealers sold 45 changers at an annual rate of 1,500,000 per year, while sales of the records rose faster than those for 78s or LPs. For the new year, Victor dropped the price of the player and phonograph to $12.95 and $29.95, while it installed the changer in a set of consoles and combinations that, at the top of the line, included television, AM and FM radio, and a changer for the other speeds.[243]

Prototype advertisement sent to RCA managers showing college student enjoying a 45J attachment. *Courtesy Camden County Historical Society.*

Prototype advertisement sent to RCA managers showing spinning records and 45J attachment. *Courtesy Camden County Historical Society.*

Prototype advertisement sent to RCA managers showing lady dancing to 45J attachment connected to RCA Victor radio. *Courtesy Camden County Historical Society.*

The label also began recording performances for its own longplaying records. Its classical performers had watched their royalty income shrink and as their contracts came up for renewal, they demanded a Victor LP. Victor's leadership agreed, but extracted endorsements of the 45 in return. In the one advertisement placed in *The Review*, none of the artists declaiming on the fidelity of the 45 were featured in the albums listed for sale.[244]

Accepting the Inevitable: RCA Victor's Longplaying Record, January 1950

With the momentum of the Christmas season sales and the prospect of continued heavy marketing, RCA Victor and its customers assured the future of the 45. On January 4, 1950, RCA Victor announced the imminent arrival of its own longplaying disc. In a full-page ad in the *Review's* "Recordings" section, the "pioneer and world leader in recorded music" announced a new policy over Frank Folsom's signature. In the future the "World's Greatest Artists and Music" would appear on "ALL phonograph speeds!" The copy still focused on the 45, which was "here to stay and is destined to lead all other types of recorded music." In a bid for the heartstrings of older music lovers, RCA Victor restored the "Victrola" trademark in a line of home entertainment consoles that could play all three speeds. The label also promised to issue all new releases on 78 "as long as there is a demand for them." Finally, "to serve those music lovers who wish to play certain classical selections on long-playing records," Victor would begin offering "a new and improved" record using "an exclusive RCA Victor processing method which ensures high quality and tonal fidelity."[245]

The music lovers and the industry applauded the move. Irving Kolodin noted with satisfaction that he had given Victor one year to realize that the "smaller if more discriminating" segment of consumers would not accept the 45 for classical performances.[246] Still, with a new administration at RCA Victor making every effort to mollify the music lovers, Kolodin found himself praising the label's LPs as "better than I remembered them from 78 listening."[247] C. G. Burke thought it "not improbable that Victor, which has fewer discs substandard technically than any other company," would immediately produce LPs of high quality. The reviews that followed the first releases in April justified his prediction and ensured the LP's status as a vehicle for high culture and high fidelity.[248]

Double page ad from Oct. 1950 issue of *Radio and Television Journal* featuring Jack Paar. *Courtesy Camden County Historical Society.*

Chapter Three

Evolution and Improvement of the "45"

Center Hole and Label Modifications

Very early 45 rpm records had what came to be called a "notch," which was an abrupt reduction in the record's thickness, very close to the center hole. With the rest of the label at "normal" thickness, this reduced thickness at the center hole provided spaces between the records when they stacked upon each other so that the record-dropping mechanism did not have to force its way between the records. In the spring of 1949 this was changed from a notch to a gradual sloping of the record thickness towards the center hole. It's not clear why this change was made. There was a problem with records slipping while playing but this change yields *less* label surface, not more. In the fall of 1949 RCA changed the composition of the record label to a less slick material. This did resolve the slippage issue. In the spring of 1950 the end groove was modified to more abruptly bring the tonearm to the reject point.

Early notched 45 rpm record.

In the spring of 1949 the notch was changed to a gradual reduction in thickness towards the center.

Colored Vinyl Discarded

There is no doubt that the differently colored vinyl 45 rpm records were a nice visual touch and that the colors provided assistance with inventory control. It was soon evident, however, that other manufacturers making black records could include impurities in their vinyl, thereby cutting the cost of each record. Colored records had to be made of pure vinyl with no impurities; otherwise the impurities would show in the semi-transparent record surface. In order to conserve costs, RCA Victor began using different colored labels instead of colored vinyl as early as late 1950. The only exceptions were the yellow children's records, which remained yellow until 1952, and the red classical records, which remained red until 1953.

Label Updates

RCA Victor's black vinyl '45' went through the following label updates:

- 1950: Hard to read gold lettering changed to silver
- 1950: The number 45 starts appearing in boxed form
- 1952: Blue label changed to black.
- 1953: The words "new orthophonic" added to the label.
- 1954: New layout featuring Nipper logo in color on the top
- 1968: Orange label with new layout

Additional changes were made after this but since this book deals with a limited time period the above should suffice.

Group of three RCA Victor 45s showing early blue label, later black label, and the most famous and popular label with Nipper logo in color.

Fidelity Improved

Early 45 rpm records seemed to have low volume compared to those that were manufactured later. Constant improvements in microphones and recording equipment technology during the 1950s permitted both greater volume and greater dynamic range on the records. Another improvement in the record manufacturing process allowed the groove being cut in the record to be size-adjusted depending upon the loudness of the music passage being recorded. In other words, soft passages could have the grooves cut closer together than those cut for loud passages. This allowed more music information to fit on each disc and improved the dynamics of the sound.

The Extended Play (EP) 45 RPM Record

In 1952 RCA Victor introduced the extended play (EP) 45 rpm record. It could hold nearly twice as much music as the conventional 45 but was priced only slightly higher. Extended play records were promoted heavily as an answer to Columbia's LP record, but they did not supply complete satisfaction. It was still necessary to play three or four records to equal the content of one LP, and the volume and fidelity of the 45 EP records were compromised by the grooves being scrunched together.

The *Extended Play* 45 typically contained two songs per side.

Examples of early 45 *Extended Play* (EP) album boxes.

Subsidiary Labels

To try to reach a larger audience and market share, RCA Victor created the following subsidiary labels:

- RCA Camden: 1955 to 1970s
- Groove: 1953 through 1958
- RCA Bluebird: 1953 and 1954

Two examples of RCA subsidiary *Camden* records.

Example of an RCA subsidiary *Groove* record.

Example of an RCA *Bluebird* series record.

Other "45" Facts

- Stereo 45s were made in 1958 by RCA Victor and Roulette records. They did not sell well and did not reappear until the late 1960s.

- Many of the early release records made for disc jockeys (promos) would have the same song on both sides, one side would be stereo and the other would be monophonic.

- Colored vinyl has recently become popular again. The generation of "Baby Boomers," who are filling up their own vintage jukeboxes, are buying them from specialty companies.

Colored vinyl has become popular again on a limited basis. It is available today from certain companies.

Stereo 45s released in 1958.

Chapter Four

1949: Product Introduction

January 10, 1949 was the day that RCA Victor introduced its new system of recorded music to the public. Articles and advertisements appeared in many of the newspapers and magazines of the day, including *The New York Times*. The public was told that the hardware would be available in stores on April 1, but it actually arrived a day early, enabling stores to begin selling the new system on March 31. There was a model for every lifestyle: those with inexpensive attachments were at one end of the spectrum and, on the other end of the model line, were those with beautiful wood consoles featuring AM/FM radio, two record changers, and eyewitness television.[249] Cabinets of Bakelite, wood, and vinyl covered wood were available. A few special units were housed in Plexiglas so the dealer could demonstrate the record changer's inner workings for prospective customers.

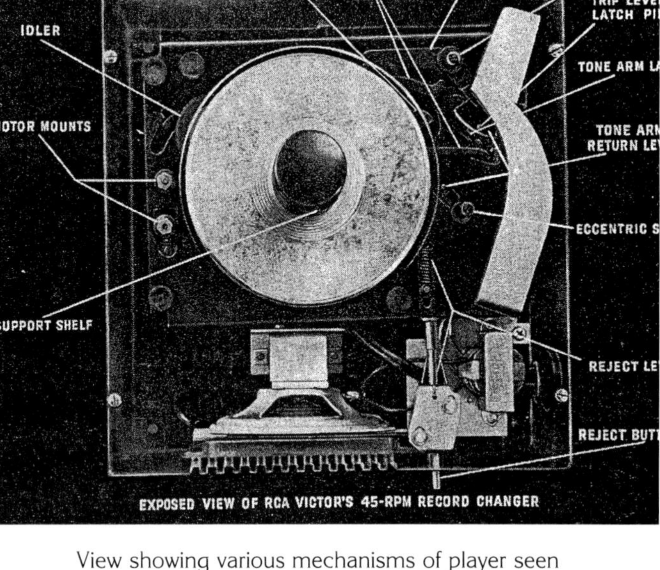

View showing various mechanisms of player seen through Plexiglas cabinet.

The "NEW LOOK" In Popular Records

Closeup of the new RCA record player and its 45 r.p.m. record. The 7" vinyl plastic record has 1½ inch center hole.

By

TOM GOOTÉE

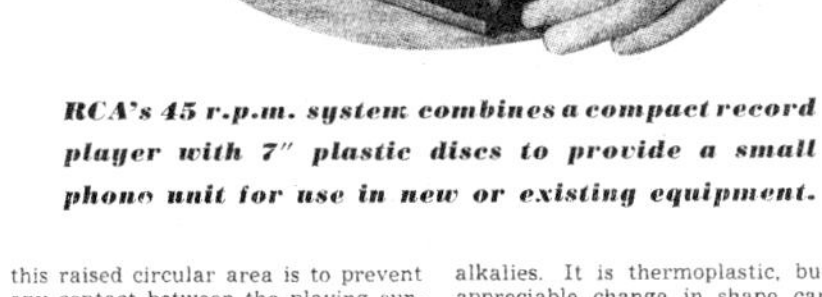

A NEW and important trend toward high fidelity and the distortion-free reproduction of recorded music and entertainment in the home is indicated by the radically new system of 45 r.p.m. records and matched record players developed by the *RCA Victor Division* of *Radio Corporation of America*.

Establishing new standards of size and speed as well as improved fidelity, the 45 r.p.m. system is designed to provide mechanical simplicity, small size, light weight, and lowered costs.

The Records

The new 45 r.p.m. records are wafer-thin, non-breakable discs of the vinyl plastic, which is known commercially as *Vinylite*. All records are of uniform size—slightly less than seven inches in diameter. All of the records have a large center spindle hole which measures one and one-half inches in diameter.

The playing surface on each side of a record is confined to a single band, about one inch in width (maximum), which represents a maximum of 275 grooves. This band represents a playing time of about five minutes. A three-minute record would have a narrower band of grooves and correspondingly fewer grooves.

Between the band, or playing surface, and the large spindle hole is a slightly raised collar which carries the record label. The primary purpose of this raised circular area is to prevent any contact between the playing surfaces of proximate records when they are stacked together. In this way, scratches due to friction with other records is effectively minimized.

The *Vinylite* used in the records is a hard and durable material and stands considerable abuse. Essentially a vinyl-acetate resin, *Vinylite* is molded with heat and pressure like other synthetic resins. *Vinylite* is unaffected by water, oil, gasoline, acids, or alkalies. It is thermoplastic, but no appreciable change in shape can be detected when the record is subjected to normal heat. The material can be produced in a variety of colors and because of its high refractive index the colors are brilliant and clear.

Illustrative of the relatively small size of these new records is the fact that a stack of 1000 discs can be housed in an ordinary console cabinet. Thus the problem of record storage in the

(*Continued on page* 98)

RCA's 45 r.p.m. system combines a compact record player with 7" plastic discs to provide a small phono unit for use in new or existing equipment.

Comparison in size between the 78 r.p.m. and the new 45 r.p.m. discs. The new 7" vinyl record plays 5 min., 15 sec. per side.

Over-all view of the record player in operation. The new unit can be used with any type of available audio system.

March, 1949

This magazine article from March 1949 discusses the new record and record playing system from RCA Victor.

VOL. 2, No. 14 APRIL 18, 1949 RELEASE 49-14

DOZEN POP HITS LISTED ON 45RPM

Top Sellers Will Be Out On New Discs In Next Few Weeks

To stimulate sales of the new 45 system a dozen tunes riding high in the hit parade will be pressed in the next few weeks on transparent, seven-inch vinyl and labeled black.

The discs will herald the first up-to-the-minute cuttings on 45 of what's tops in pops—the stuff they can't get along without. Next month will see the entire pop release made available on both 45 and 78. The practice will continue regularly thereafter.

Release of the current successes is an additional boost to what already has been put out to help carry 45 in the popular department—25 albums and 11 singles. A heavier backlog is being built up speedily. The leading hits and 45 catalog numbers are:

(1) "A—You're Adorable" and "When Is Sometime?"—Perry Como (47-2899).

(2) "Dreamer With A Penny" and "I'm Beginning To Miss You"—Bill Lawrence (47-2893).

(3) "Careless Hands" and "Powder Your Face With Sunshine"—Sammy Kaye (47-2962).

(4) "Red Roses For A Blue Lady" and "The Melancholy Minstrel"—Vaughn Monroe (47-2889).

(5) "She's A Home Girl" and "Enjoy Yourself"—Tommy Dorsey ([illegible]).

(6) "Lullaby In Rhythm" and "Birdland"—Charlie Ventura (47-2891).

(7) "Ya Wanna Buy A Bunny?" and "Knock, Knock"—Spike Jones (47-2894).

(8) "Hurry! Hurry! Hurry!" and "Ballin' The Jack"—Three Suns (47-2898).

(9) "Some Enchanted Evening" and "Bali Ha'i"—Perry Como ([illegible]); "I'm Gonna Wash That Guy Right Out A My Hair" and "A Wonderful Guy"—Fran Warren (47-2897); "Don't Rob Another Man's Castle" and "There's Not A Thing"—Eddy Arnold (48-0042); "Forever And Ever" and "I Don't See Me In Your Eyes Anymore"—Perry Como (47-2852).

"RECORD REVIEW" IS BIG SUCCESS

Circulation Of New Magazine Skyrockets As Praise Pours In

Statistics haven't yet been got up, but the success of the first issue of the colorful RCA Victor Record Review combined with In The Groove has been immediate and, for a consumer publication, phenomenal. Copies were snapped up throughout the country, no one section showing more interest than another. Congratulatory letters poured in.

The suddenness of the success is being attributed partly to the stature of the writers in the new mag, who for the most part are celebrities in the musical and allied spheres. In forthcoming issues even more "names" will be commissioned to write under the colors of RCA Victor.

record review

The second number (May), which will be in hand by the time you read this, has pieces-done-up-in-art by Ray McKinley, Burt Lancaster, Connie Mack, Claude Thornhill, James Melton, Tex Beneke, Jan Peerce, Philip Wylie, Al Nevins, and so on. The art accompanying these is stunning.

Besides name writers another attention-getting device will be articles that sharpen musical intelligence, like "Story Of A Rodgers And Hammerstein Score," planned for the third issue. More, there will be a picture story on what a star undergoes on a club date.

The Record Review is a brilliant innovation from the standpoint of sales. It sells without ever praising a record or asking the reader to buy one. There is no badgering by advertising of the readers of this publication.

First page articles from *Record Bulletin* for April 18, 1949. *Courtesy Bill Pouluh Collection.*

Model 9JY: The First 45 RPM Phonograph Attachment

The first 45 rpm phonograph attachment, model 9JY, featured the new rp-168 record changer. RCA Victor's Chief Engineer, Benjamin Carson, used the prototype he designed in the early 1940s and updated it with a new tonearm and cabinet. The turntable was brass and built into the large center spindle was a system of rotating shelves and separators, which allowed records to drop onto the turntable one at a time. These Bakelite attachments were designed to be plugged into an external amplifier found in radios and televisions of the era. RCA Victor had been equipping their products with a phono jack and switch so that the new record player could be used with any RCA Victor radio or television. If one's radio or television set did not have such a jack, it could be added by a technician. This was an inexpensive way to introduce the new phonograph to the customer's existing system. RCA Victor also produced a special album called the *45 rpm Introductory Album*, which consisted of ten records featuring such top-notch artists as Vaughn Monroe, Perry Como, and Eddie Fisher. This collection consisted of several of the different colored vinyl records, thereby introducing the new concept to new customers.

9JY Record Changer attachment in dark brown Bakelite, shown with *45 rpm Introductory Album*. Power and volume control are provided on lower right side. Front mounted pushbutton starts reject cycle. Separate audio cord provides connection to any radio or television so equipped. Introduction-priced at $24.95 and reduced to $12.95 in September, 1949. $75-$120 (does not include album).

9JY Record Changer attachment with see through case next to conventional 9JY. These specially made units were for demonstration purposes only, showing the insides of the new record playing system. The cabinet is made of Plexiglas. *Courtesy Doug Houston Collection*. $500-$600 for see through unit.

Model 9JY dealer sheet. *Courtesy Lee Wells Collection*.

An Inexpensive Attachment to Bring You Advantages and Savings of the New RCA Victor System

RCA VICTOR 9-JY

The new RCA Victor system of recorded music offers more enjoyment and advantages than any other system. You can have this added pleasure and new, convenience inexpensively through your present set. Model 9-JY is easily connected to your set. The RCA Victor 9-JY attachment plays 45 rpm records like the new RCA Victor distortion-free record. *Only 7 inches*, it plays as long as an ordinary 12-inch record . . . costs much less. Here is the world's fastest changer . . . the easiest operating ever designed. Just place up to 10 records on the large spindle and press a button. The changer can play more than 50 minutes before requiring attention. And you hear only the music you want when you want it. You can build a library of both "*classics*" and "*pops*" at great savings. And there is no storage problem . . . more than 150 singles or 18 albums of the 7-inch records fit in one foot of bookshelf space . . . can be stored in desks and tables. The 9-JY fits the storage compartment of some consoles. This is truly the modern, inexpensive way to enjoy recorded music.

In this prototype ad, Fran Allison, popular member of the *Kukla, Fran and Ollie* 1950s TV show, works on "45" commercials with Cecil Bill and Fletcher Rabbit, two of the "Kuklapolitan" players. *Courtesy Camden County Historical Society*.

9EY3 Table Top Phonograph

RCA Victor's first 45 rpm table phonograph used the new rp-168 record changer. RCA Victor heavily advertised the "golden throat" tonal quality of their amplifiers and speakers. All 1949 models were connectorized[250] for easier servicing. Very early in the first production run, the decorative louvers went all the way down to the bottom of the front panel of the Bakelite case, as shown. Because these phonographs were so easy to handle, dealers were encouraged to sell one to each member of a customer's family. An optional tan carrying case was available, thereby turning the unit into a portable record player.

9EY3 table model phonograph in dark brown Bakelite case along with one of the new RCA Victor colorful album sets. Power and volume control are provided on lower right side. Front mounted pushbutton starts reject cycle. Optional soft carrying case also available at extra cost. Introduction-priced at $39.95 and reduced to$ 24.95 in September, 1949. $175-$275.

Model 9EY3 featured on front cover of *Radio Age* magazine, April 1949.

Model 9EY3 dealer sheet. *Courtesy Lee Wells Collection.*

In another prototype ad, Burr Tillstrom, creator of the "Kuklapolitans," looks over a model 9EY3 with "salesmen" Kukla, Ollie, and Fran Allison. *Courtesy Camden County Historical Society.*

9EY31 and 9EY32 Portable Phonograph

The first portable 45 rpm phonograph with amplifier and speaker, models 9EY31 and 9EY32 were available in wooden cabinets covered in red (31) or brown (32) vinyl. These units were much more rugged than their Bakelite cousins. They also sounded better because their cabinets housed a larger speaker. These models did not, however, sell nearly as well as the Bakelite models and it is uncertain whether their appearance or their higher cost accounted for the slow sales.

Prototype advertisement sent to RCA managers for portable model 9EY31 and 9EY32. *Courtesy Camden County Historical Society.*

Model 9EY31 and 9EY32. First portable 45 rpm phonograph with amplifier and speaker. These models were available in tan or red vinyl. Volume and tone controls are front mounted on each side of speaker grill. Originally priced at $49.95. $125-$200.

A Wonderful Musical Companion With All the Advantages of the RCA Victor "45" System

RCA VICTOR 9-EY-32

The RCA Victor 45 rpm system has been acclaimed as the modern, inexpensive way to enjoy recorded music. This history-making development has given music lovers advantages and conveniences never before found in any method of playing records. Now, RCA Victor offers you these advantages and conveniences ... this superb listening pleasure ... in a smart portable. Model 9-EY-32 brings you such modern features as the world's fastest record changer ... the easiest, surest operating changer ever designed. Just plug it into any AC outlet ... it can play up to ten of your favorite selections on easy-to-handle, low-cost, 7-inch records ... up to 50 minutes of music at one touch of one button. These vinyl plastic records can play as long as ordinary 12-inch records. The music you hear is distortion-free. And you always enjoy the vibrant voice of the "Golden Throat" tone system. A smartly styled case of simulated leather will always be an attractive traveling companion. More fun than you dreamed possible will be your reward when you take Model 9-EY-32 along.

"Golden Throat" TONE SYSTEM

Dealer sheet for model 9EY31 and 9EY32. *Courtesy Lee Wells Collection.*

Children's Phonographs

These are the first "cartoon players" manufactured by RCA Victor. The unit is basically a 9EY3 with the dark Bakelite case painted white and decals added. Since paint from this era was lead based, there was a cautionary note on the bottom of the cabinet regarding children and the dangers of lead poisoning. It is ironic that the only lead-painted phonographs were for the children! Two models were available—one bore the image of cowboy singer/actor Roy Rogers and one featured characters from the Walt Disney cartoons. Interestingly, the latest popular characters, like Mickey Mouse, Donald Duck, Goofy, and so on, did not appear. Characters featured on these early "cartoon players" were Pinocchio, Snow White, the Seven Dwarfs, Johnny Appleseed, etc. These units were available for only one year, so today they are quite rare; this is especially true for the Roy Rogers model.

9EY35 Children's Record Player (Disney). Bakelite table model painted white with Disney decals added. Volume control is on the lower right side. Some advertisements for this model show a different decal arrangement. Originally priced at $39.95. $200-$325.

Prototype advertisement sent to RCA managers for Disney model 9EY35. *Courtesy Camden County Historical Society.*

Ad from trade magazine featuring 9JY and 9EY36 models. *Courtesy Camden County Historical Society.*

9EY36 Children's Record Player (Roy Rogers). Bakelite table model painted white with Roy Rogers decals added. Volume control is on the lower right side. Originally priced at $39.95. $250-$400.

You can offer your customers more for their money...

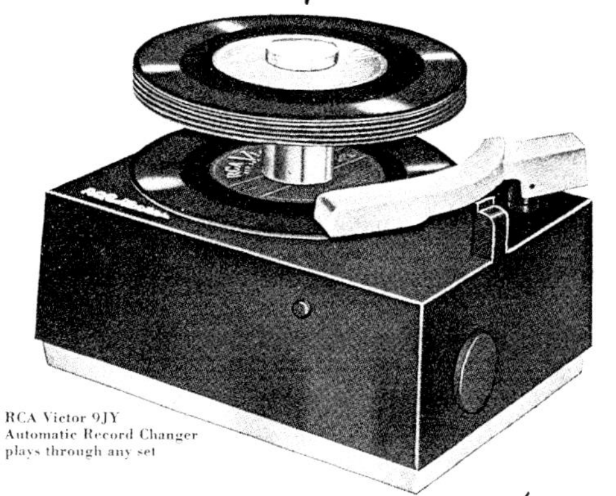

when you sell the 9JY

FOR ONLY $12.95*

*suggested list price

Here's a real traffic-builder. It has the lowest price in history for a *completely automatic* record changer. A value that *draws* customers into your store. Your selling job is made *easier* and *simpler*. Extra traffic increases your prospects for the appliances and other items you're featuring, too.

Promoting Model 9JY ties you in neatly with RCA VICTOR'S smashing, nationwide promotion—identifies your store as headquarters for what's *new* and *progressive!*

when you sell the 9EY36

*suggested list price

"Santa Claus" himself will shop your store for *this* great value with "Christmas-present" appeal. The most nearly *child-proof* automatic record changer ever offered anywhere, anytime! Appeals to children *and* parents. Walt Disney, Model 9EY35, or Roy Rogers, Model 9EY36 design.

Every time you sell these "45's", you're set for $10 to $15 worth of record sales. These customers are good prospects for your appliances, too. Boost your reputation for high quality at modest prices —with the RCA VICTOR "45's"!

Sell the entertainment systems of the future TODAY RCA VICTOR records, players, radio and television sets!—

TT-6106

Roy Rogers model 9EY36 dealer sheet. *Courtesy Jon Butz Fascina.*

The Old West and the New "45" Meet In a Colorful Roy Rogers Phonograph

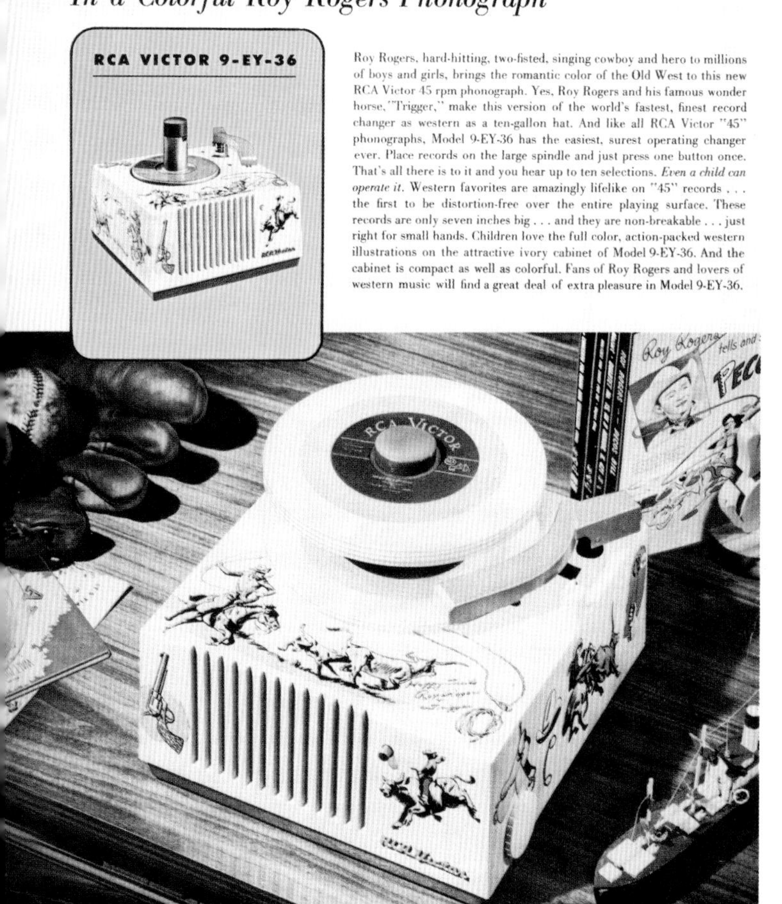

RCA VICTOR 9-EY-36

Roy Rogers, hard-hitting, two-fisted, singing cowboy and hero to millions of boys and girls, brings the romantic color of the Old West to this new RCA Victor 45 rpm phonograph. Yes, Roy Rogers and his famous wonder horse, "Trigger," make this version of the world's fastest, finest record changer as western as a ten-gallon hat. And like all RCA Victor "45" phonographs, Model 9-EY-36 has the easiest, surest operating changer ever. Place records on the large spindle and just press one button once. That's all there is to it and you hear up to ten selections. *Even a child can operate it.* Western favorites are amazingly lifelike on "45" records . . . the first to be distortion-free over the entire playing surface. These records are only seven inches big . . . and they are non-breakable . . . just right for small hands. Children love the full color, action-packed western illustrations on the attractive ivory cabinet of Model 9-EY-36. And the cabinet is compact as well as colorful. Fans of Roy Rogers and lovers of western music will find a great deal of extra pleasure in Model 9-EY-36.

Ad for Little Nipper Children's Records featuring Roy Rogers model.

The First Radio/Phonograph Models

Consumers had a choice of wood (9Y7) or Bakelite (9Y51) models. The wooden model was available in light and dark wood finishes and provided some storage space to hold records on each side of the record changer. Bakelite was available only in dark brown.

Model 9Y7 Phonograph and radio with front mounted slide rule dial and wooden cabinet. This table model was available in mahogany and oak finishes. Volume and tuning controls on lower front. Tone selector is mounted on lower left side. Originally priced at $79.95 for mahogany, $89.95 for oak. $125-$200.

Model 9Y51 Radio and Phonograph in dark brown Bakelite. Very popular table model with tuning and volume thumb wheels on each side of the slide rule dial on the front. Tone selector is mounted on lower left side. Originally priced at $59.95. $200-$300.

Dealer sheet for model 9Y7 Radio Phonograph. *Courtesy Lee Wells Collection.*

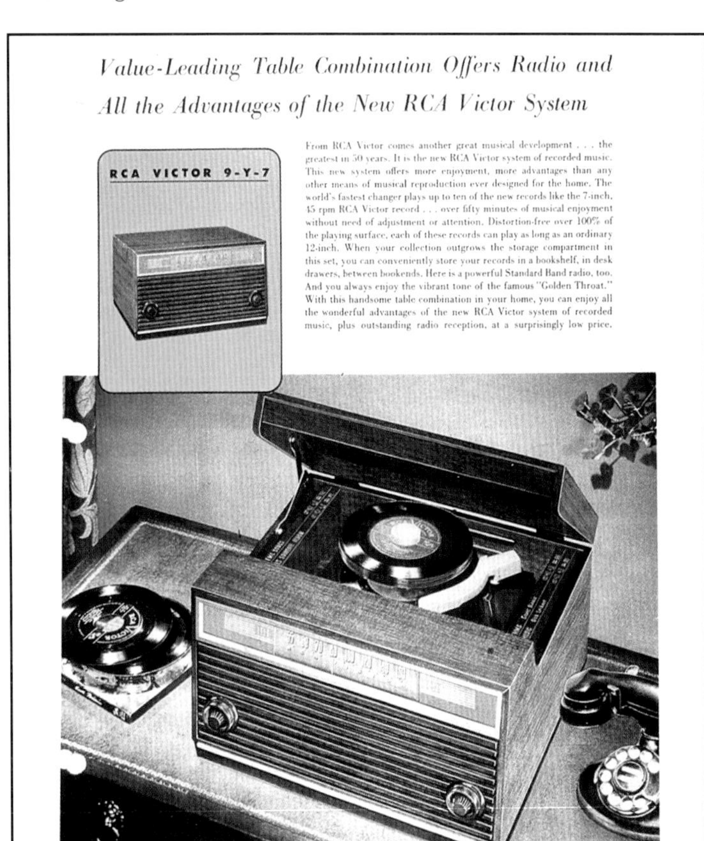

Dealer sheet for model 9Y51. *Courtesy Lee Wells Collection.*

A separate wooden model 9Y5 Radio and Phonograph was available in Canada only. This wooden table model was completely different than the American model 9Y7. The cabinet is smaller and more rounded, with the radio slide rule dial angled up in the front. Volume and station selection are controlled by side-mounted knobs, while tone and input selection are controlled by front-mounted knobs. This model was available in AM or AM/SW. The model shown is AM only.

Model 9Y5 Radio and Phonograph (only available in Canada). This wooden table model has volume and station selection as side mounted controls, while tone and input selection are front mounted. It was available in AM or AM/SW. The model shown is AM only. Original price $99.95. $150-$250.

Canadian ad showing the new RCA Victor system of recorded music. Notice the higher prices compared to those of the contiguous 48 states.

The Berkshire Phono Attachment, Model CP 5203

This was the most expensive and quality made attachment by RCA Victor, featuring a beautiful cherry wood cabinet and brass hinges and lock. It was designed to plug into the very prestigious and expensive Berkshire High Fidelity Console that sold for $4000 in 1948. It is the only 45 rpm player made in the USA with a magnetic cartridge and special capacitive-start motor with built-in cooling fan. A special black color scheme was used on the motorboard and tonearm. The turntable pad was black felt.

Color Canadian ad for RCA Victor phonographs including model 9Y5.

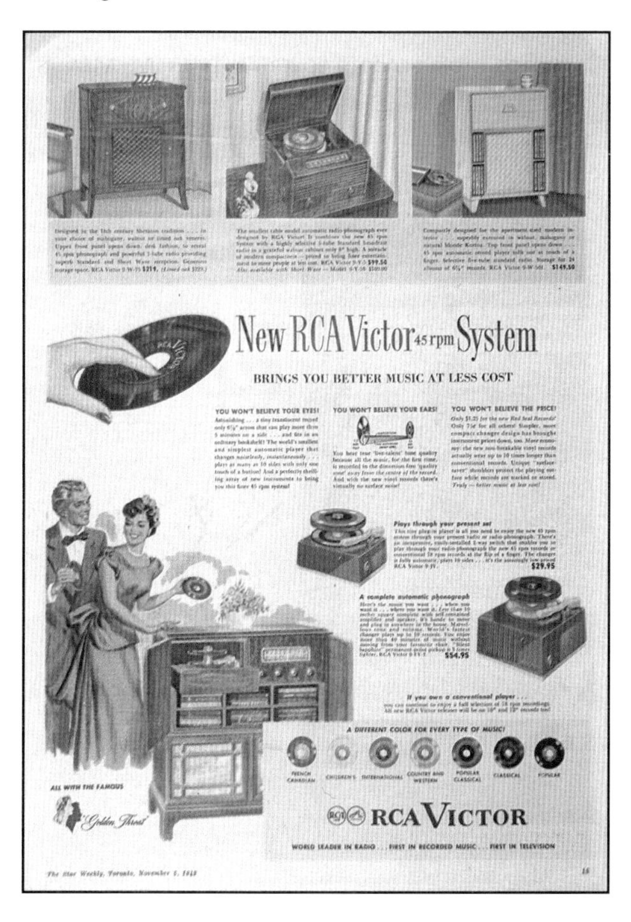

RCA Victor brochure showing many of the 1949 models.

The Berkshire Phono attachment model CP 5203. Table model featuring beautiful cherry wood cabinet and brass hinges and lock. Special black color scheme was used on motorboard and tonearm; turntable pad is black felt. Original price unknown. $250-$350.

The Berkshire Phono attachment model CP 5203 shown with closed lid.

Magazine ad illustrating the new fast acting record changer.

Small Unusual Consoles

Because of the compactness of the new 45 rpm system, RCA Victor designed some models that highlighted the players' unobtrusiveness. The 9W51 looked like an end table and was equipped with a 45 rpm phonograph in a slide-out drawer and a radio with an eight-inch speaker mounted above the drawer. Model 9W102 was a coffee table with built-in radio/phonograph.

Ad from *Radio and Television Journal*, September 1949, showed merchandising possibilities.

Model 9W51 Compact Console Radio and Record Player. Low, medium, or high tone is switched in by one of two knobs on the left side of the cabinet. The other knob on the left provides volume control. A single knob on the right side provides tuning of the radio. Ample storage is provided under the sliding record changer drawer for 45 rpm records. A large eight-inch speaker is mounted near the top, in the front of the wood cabinet, providing a nice sound quality. 9W51 originally priced at $99.95. $175-$275.

Diminutive Radio-Phonograph Styling

Outstanding Listening Pleasure

RCA VICTOR 9-W-51

Limited space is no longer a factor when planning home entertainment . . . not when you choose the Victrola radio-phonograph 9-W-51. This instrument, designed especially for homes where space is at a premium, is diminutive in size alone. In quality and performance, it towers above many more expensive radio-phonograph consoles. The 9-W-51 offers you the incomparable pleasures of the RCA Victor 45 rpm system and the enjoyment of Standard Band radio. It has the world's fastest, finest record changer . . . the easiest, surest operating changer ever designed. Plays up to ten distortion-free, non-breakable records at one press of one button. You enjoy the full-rich tone of the "Golden Throat" for both radio and record listening. In walnut, mahogany or limed oak finish, Model 9-W-51 has the appearance of a compact chest or table. This versatile set is an excellent example of the unique styling made possible by the "45" system. Model 9-W-51 affords finest listening pleasure and topmost quality at a truly modest cost. "Victrola"—T.M. Reg. U.S. Pat. Off.

Dealer sheet for model 9W51. *Courtesy Lee Wells Collection.*

Model 9W51 Compact Console Radio and Record Player shown with changer drawer closed. Ample storage is provided under the sliding record changer drawer for 45 rpm records.

Dealer sheet for model 9W102 wooden coffee table record player. The record changer is accessible through a slide out drawer. Very rare model; I have never had the privilege of hearing or owning one. Original price unknown. $250-$350 for actual set. *Courtesy Lee Wells Collection.*

Victrola Triumph! Coffee Table Radio-Phonograph

AM-FM Radio—RCA Victor 45 rpm Phonograph, 12-inch Speaker

RCA VICTOR 9-W-102

A truly enchanting departure in radio-phonographs, the Victrola 9-W-102. This handsome combination is at once a modern, functional coffee table and Victrola instrument. With it you enjoy the vast radio pleasures of Standard Band and RCA Victor FM, Frequency Modulation at its finest. For record pleasure . . . the world's finest records and the world's fastest changer . . . the RCA Victor 45 rpm system of musical reproduction. Slip up to ten 7-inch, distortion-free records on the big spindle . . . press a button once and enjoy a "live talent" concert. And Model 9-W-102 affords luxurious operating convenience. As you open the drawer on the right, the changer and the control panel for both radio and records come forth. The door on the left opens into storage for up to 216 records. Naturally, there is the magnificently rich voice of the "Golden Throat." A 12-inch speaker! This distinguished instrument is available in handsome walnut, lovely mahogany or modern limed oak finish. It is for those persons who set the pace for luxurious living.

Consoles

RCA Victor produced many different consoles featuring various combinations of AM radio, FM radio, two separate record changers (one for the 45 system and one for the music enthusiast's collection of 78 rpm records), and eyewitness television. It is curious that while other manufactures designed multispeed changers to handle the different records, RCA Victor marketed two separate record changers for several years, hoping that eventually the 45-system would be the accepted one. Ultimately RCA Victor succumbed, and in 1952 the company began installing multispeed changers in its consoles.

Victrola Console with 18th Century Charm—

45 rpm System, 78 rpm Changer, AM-FM Radio

RCA VICTOR 9-W-106

Eighteenth century cabinetmakers, acclaimed masters of their art, created furniture pieces that are today revered as priceless museum treasures. This beautiful new RCA Victor console is a replica of one of the 18th century masterpieces. But the warm, antique styling of this console is not alone its artistic triumph. Here is masterful music. The magic of RCA Victor's "Golden Throat" reproduces recorded music so faithfully that one could close his eyes and well imagine the artist performing before him. From the wonderful, distortion-free RCA Victor seven-inch, 45 rpm records; from ordinary ten- and twelve-inch records; from the powerful RCA Victor AM-FM radio; from all these comes a full, rich, vibrant tone. For as long as fifty minutes, the RCA Victor 45 rpm changer can conduct a favorite symphony or a popular jazz concert. On a separate changer, ordinary records are handled automatically, too. Discriminating people, connoisseurs of true beauty, will make this magnificent console radio-phonograph the center of their home.

"Victrola"—T.M. Reg. U. S. Pat. Off.

Dealer sheet for model 9W106 console. Many of RCA Victor's consoles came equipped with two record changers, one to play the new 45 rpm records and another to play a current collection of 78 rpm records. Original price unknown. $125-$175 for actual set. *Courtesy Lee Wells Collection.*

Complete Musical Enjoyment—Rich Traditional Styling

RCA Victor "45" System, 78 RPM Record Player and AM-FM Radio

RCA VICTOR 9-W-78

Only RCA Victor could design a console instrument with so many musical advantages ... so much unusual charm ... at such a modest cost. Here are two record changers, a powerful AM-FM radio and ample storage space. For your present collection of ordinary 78 rpm records, an automatic RCA Victor changer slides forth, at a touch, to play twelve 10-inch or ten 12-inch records. In another drawer is RCA Victor's marvelous "45", the world's fastest record changer, ready to play up to ten of the new 7-inch distortion-free records. And as you acquire a collection of these wonderful, new, low-cost "45" records, you can store them within this console. There is room for as many as 24 albums or 196 singles. The "Golden Throat" tone system provides brilliant, enjoyable listening ... through a large, twelve-inch speaker. And what an enviable addition to your home this console will make. Styled in the Traditional manner it will add richness and dignity to the loveliest interiors. Surely everyone who desires the broadest range of fine musical entertainment will find this new Victrola console a superb choice. "Victrola"—T. M. Reg. U. S. Pat. Off.

Dealer sheet for model 9W78 console. This was another of RCA Victor's consoles that came equipped with two record changers, one to play 45's and one to play 78's. Original price unknown. $125-$175 for actual set. *Courtesy Lee Wells Collection.*

Colorful double page ad introducing the new RCA Victor records and players.

"Enjoy the music you want when you want it." One of RCA Victor's slogans headlines this magazine ad.

Dealer ad showing many of the new phonographs. *Courtesy Camden County Historical Society.*

Colorful ad talks about the new RCA Victor system of recorded music.

Color ad featuring the new colored vinyl records.

Not all submitted ads were used by RCA Victor. The "center pylon" referred to here was changed to "center spindle." *Courtesy Camden County Historical Society.*

Making the new 45 rpm Records and Record Players

1 Sheets of translucent plastic are reduced to proper thickness under the heated rolls of this huge machine.

2 A 45-rpm record complete with labels is removed from one of the many powerful presses in the Indianapolis plant.

1 Maximum performance of the 45-rpm record player is assured by testing its operation with a sensitive oscilloscope.

2 Quantity production is achieved on the assembly and testing lines for the new 45-rpm record players at the RCA plant in Indianapolis.

THESE scenes from the Indianapolis, Ind., plant of the RCA Victor Division reveal some of the steps that are followed in producing increasing quantities of the new, colorful, high-fidelity recordings, only 6⅞ inches in diameter, and the fast, quiet, trouble-free record-changing mechanisms.

3 Excess plastic material, called "flash," is stripped from the outer rim of the recording following the pressing process.

4 Each of these semi-automatic machines punches out the record's large center hole at the rate of 30 discs a minute.

3 Record players undergo a listening test, one of the final steps in the manufacture of the high-fidelity instruments.

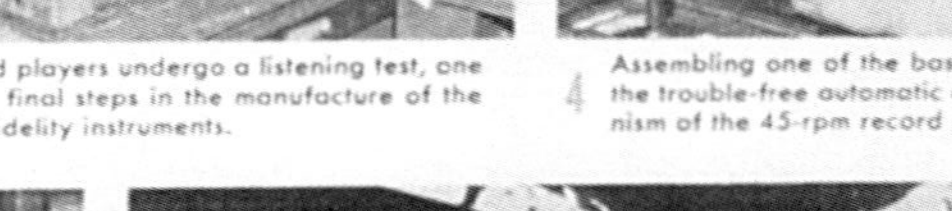

4 Assembling one of the basic components in the trouble-free automatic changing mechanism of the 45-rpm record player.

5 Exact centering of a 45-rpm record is assured by checking the movement of the outer rim under a microscope.

6 Here the finished records are placed in envelopes and then packed in containers for shipment to dealers and users.

5 An inspector checks the operation of the disc-changing cycle of a record player.

6 A moving conveyor separates the assembly group on the left from testers and inspectors of the finished product on the right.

Centerfold from April, 1949 *Radio Age*, showing the making of 45 rpm records and the new rp-168 record changer. Both were manufactured in Indianapolis. Later changer model rp190 was made in Chicago.

Chapter Five

1950: First Anniversary

In anticipation of a big sales year, RCA Victor had three different 45 record changers available during 1950. The rp-168 changer, discussed in the previous chapter, continued in production with some improvements. The pickup arm rest molded into the Bakelite case was removed and an ingenious placement of a piece of metal on the record changer chassis itself provided the same rest feature. This was a cost-cutting feature. Another cost-cutting feature was the removal of the volume control function on the attachment phonograph. The radio or television had its own volume control, so a volume control on the phono attachment was not needed. The model number 9JY was changed to 45J.

Aesthetic changes were also made. The turntable mat was now available in tan or red and the spindle cap could be red or black. The RCA Victor logo was also moved from the top of the cabinet to the front.

A 1950s family enjoying the new RCA Victor system of recorded music is shown in this RCA advertisement.

The 9JY attachment was renamed 45J in 1950. The logo was moved from the top to the front of the Bakelite case. The bottom cover, which had been painted gold in 1949, was now left brown. Turntable mats were available in tan or red and spindle caps were available in red or black. Later production used a black spindle cap that handled twelve to fourteen records in place of the earlier cap that handled ten records. Knob on lower right side turns unit on and off. Originally priced at $12.95. $75-$120.

Model 45J with original box.

Instruction booklet for 45J attachment.

RCA VICTOR 24-SHEET POSTER FEATURING THE "45" ATTACHMENT, FORM 2A5826

Printed in full color against a robin's egg-blue background, this arresting billboard plays up the quality of RCA Victor "45" through association with the famous RCA Victor trade-mark. Interest is created as passers-by see the dog, Nipper, world's best known trade-mark, turning from his familiar pose to watch the "45" record changer that's sweeping the country. As an added sales hooker, copy says "Starts at $12.95." Room left for a full 22-inch dealer imprint.

Nipper's attention turns away from the steadfast gramophone to the new 45 system. *Courtesy Lee Wells Collection.*

for a girl who has a book, a mink and an ear for the good things of life
the new
RCA VICTOR "45"
The modern, inexpensive way to play recorded music...
It's a completely automatic record changer! Plays up to 50 minutes of the music you love. And the records? They're 7-inch "book-shelf" size, non-breakable, wear up to 10 times longer. There are more than 1,200 titles to choose from... and new ones every day. It's an absolutely new idea in recorded music... the only record and changer ever designed for each other... a treasure at $12.95
7-inch Records: Red Seal 95¢*; all others 65¢*
*All prices shown are subject to change without notice. Record prices do not include Federal Excise or local taxes.
RCA Victor 9JY
Plays through your present radio or TV set
RCA VICTOR
DIVISION OF RADIO CORPORATION OF AMERICA
WORLD LEADER IN RADIO... FIRST IN RECORDED MUSIC... FIRST IN TELEVISION
DECEMBER, 1949 17

For the woman who has everything—a mink coat and a 45J.

RP-190 Record Changer, Model 45J2

The rp-190 record changer, designed in-house by Benjamin Carson, was introduced in 1950. RCA Victor contracted with the Crescent company in Chicago, Illinois to manufacture them. This model was designed with an ultra smooth rejecting cycle using a rubber cycling cam. The change cycle was no longer tied to the single rotation of the turntable. Capable of playing up to fourteen records, the turntable is made of Bakelite with a plastic spindle cap. The pickup arm is made of a very soft plastic. Both on/off and rejecting functions are incorporated into the sliding switch on the top left of the changer.

Although sales did pick up during the year, there was no need to continue with three different record changers. It was decided some time this year that the rp-190 record changer would be the one to stay in production and the others would be phased out. Rp-190 stayed in production from 1950 through 1958.

45J2 attachment. Cabinet is dark brown Bakelite. Both on/off and rejecting functions are incorporated into the sliding switch on the top left of the changer. The unit shown here has an unusual "see through" spindle cap. Originally priced at $12.95. $75-$120.

Front view of rp-190 in Benjamin Carson's patent. *Courtesy Camden County Historical Society.*

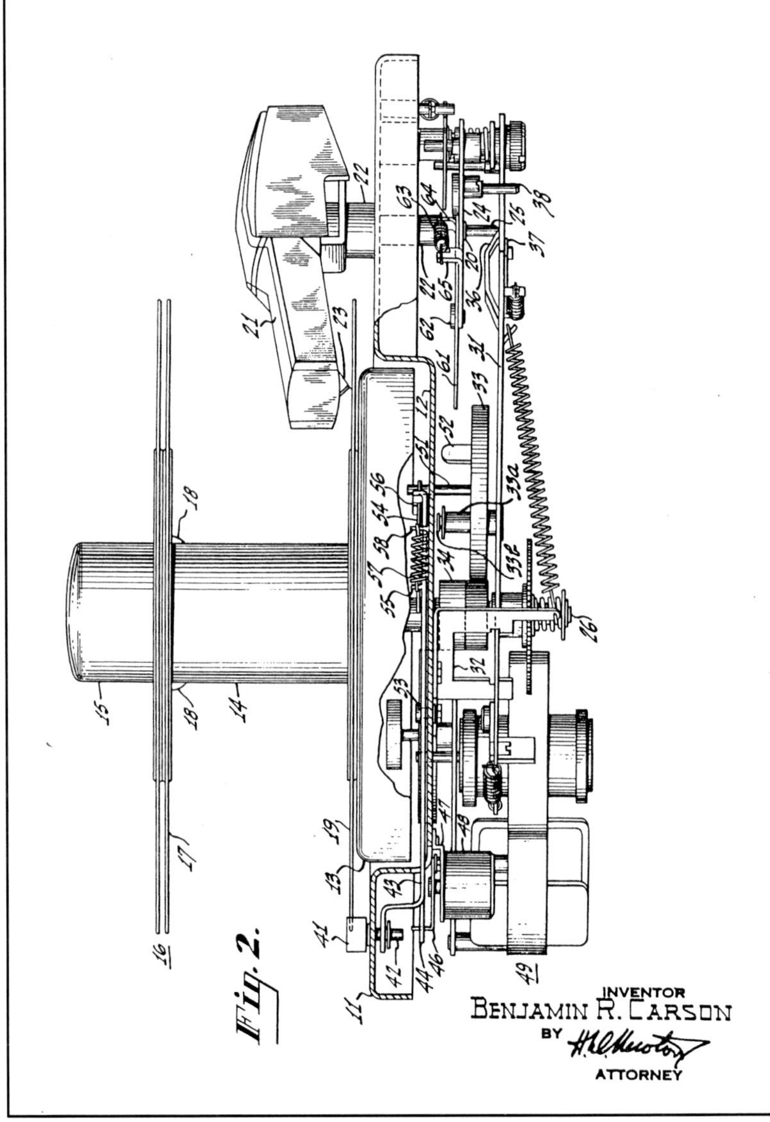

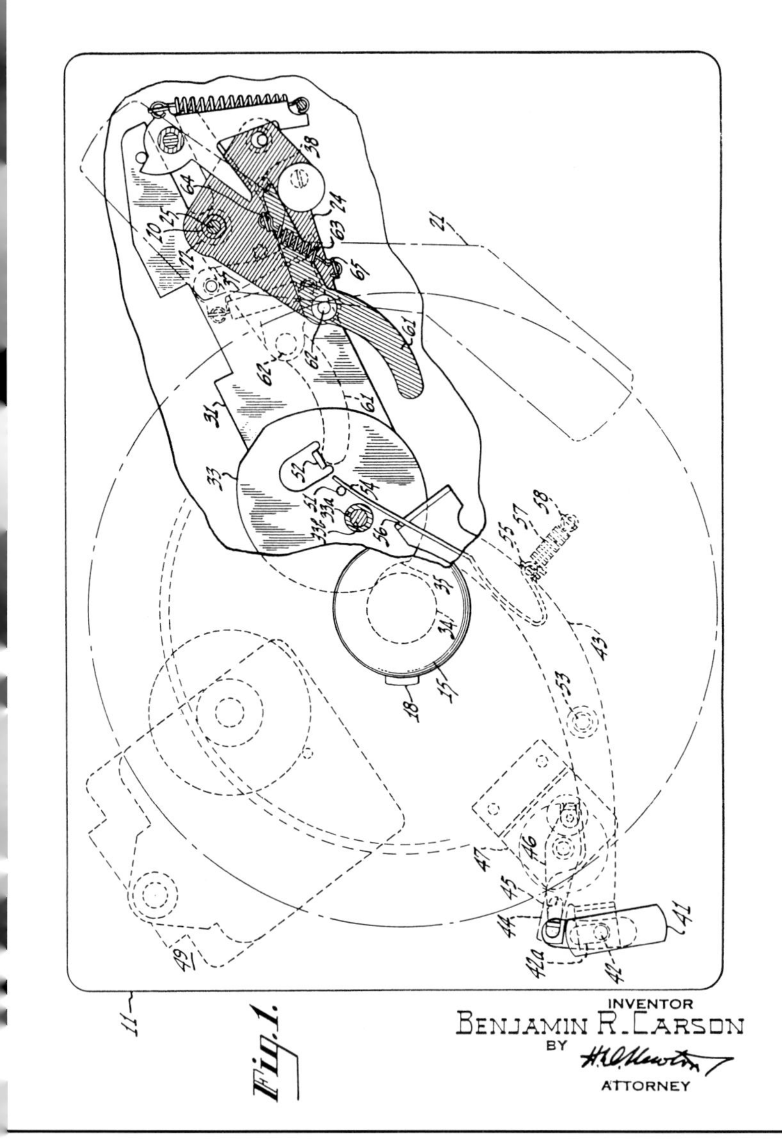

Top view of rp-190 in Benjamin Carson's patent. *Courtesy Camden County Historical Society.*

RP-193 Record Changer, Model 45J3

This record changer was available only in 1950 and was manufactured by the Oak Corporation. It was a very well made changer with a heavy metal turntable and gear drive rather than a rubber cam for the reject cycle. Apparently RCA Victor wanted to have as many 45 rpm changers available as possible in case the buying public created a high demand—hence the Oak company's manufacturing augmentation. When RCA Victor decided that it could handle the market by itself, it also decided to sell phonographs using only the rp-190 changer, and it stopped using the rp-193 changer after 1950.

The rp-193 was available only as an attachment (model 45J3) to be plugged into a radio or a television receiver. Curiously, the top of the spindle in this otherwise well-made unit is equipped with a very cheap plastic insert that wobbles around because of its poor fit. The reject cycle is quite smooth with only a hint of gear noise. The Bakelite cabinet is slightly more rounded at the corners, giving it a nice Art Deco look. The spindle was made in one piece, whereas the other changers were made in two pieces. Players incorporating the rp-193 changer are definitely scarcer than those with rp-168 or rp-190 changers, but they can still be found.

45J3 attachment. Cabinet is dark brown Bakelite featuring rounded corners that give it a nice Art Deco look. The off/on reject switch triggers the reject cycle as soon as power is applied to the turntable. Originally priced at $12.95. $100-$150.

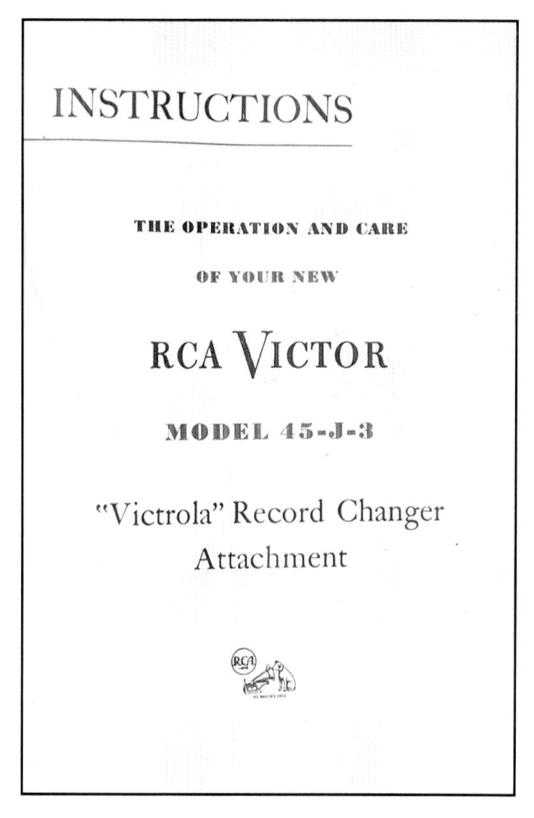

INSTRUCTIONS

THE OPERATION AND CARE

OF YOUR NEW

RCA VICTOR

MODEL 45-J-3

"Victrola" Record Changer Attachment

Instruction book for 45J3.

The 45EY Series

The 45EY series was introduced in 1950 and was available in two color combinations. Basically 1949's Bakelite model 9EY3 dressed up, the 45EY sported a two-tone paint scheme on the pickup arm and a black spindle cap. The 45EY1 had a dark maroon-colored pickup arm and red spindle cap. On the 45EY series, the top of the Bakelite case was painted gold and the RCA Victor logo moved to the top of the case.

The amplifier has two stages of audio amplification for plenty of volume when a good quality pickup cartridge is installed. Record changers were fitted with medium-output crystal pickup cartridges. The 45EY series is equipped with spring-loaded in/out type record separators within the large center spindle. Earlier ones utilized rotating separators that could jam. This new type would greatly reduce record jamming.[251]

Some 45EY units can be found with the early 9EY3 Bakelite case, which had the pickup arm rest on top. Later models did not have the arm rest on top of the cabinet. These must have been leftover cabinets from the previous year. The cases were painted gold on top, including the pickup arm rest.

45EY phonograph. Dark brown Bakelite table model with two-tone tonearm and black spindle cap, shown with brochure. Volume control is mounted on lower right side. Originally priced at $27.95. $175-$275.

45EY1 phonograph. Dark brown Bakelite table model with maroon tonearm and red spindle cap. Shown with first anniversary album. Volume control is mounted on lower right side. Originally priced at $27.95. $175-$275.

Dealer sheet for model 45EY. *Courtesy Lee Wells Collection.*

New "45" Complete Phonograph . . .
New Beauty—Greater Value!

RCA VICTOR 45-EY
WITH THE
Golden Throat
TONE SYSTEM

So very many outstanding features has this complete "45" phonograph that it is indeed difficult to select the *one* most outstanding. Countless music lovers will choose the 45-EY because of the greater enjoyment it affords. The combination of its superior performance features and the unparalleled quality of "45" recordings bring about a listening pleasure heretofore never achieved. The record is distortion-free over its entire playing surface and is heard through the famous "Golden Throat." Many persons will single out the 45-EY's ease of operation. To enjoy up to fifty minutes of music, just place up to ten records on the spindle and press one button once. Music lovers who have a keen appreciation for functional beauty will point to the 45-EY. The compact cabinet is of deep maroon enhanced by a golden colored top. Its styling is fresh and modern . . . perfectly at home wherever it is placed. And certainly, the matter of price is noteworthy, too. Model 45-EY sets a new standard of value for a system whose brilliant values have made it so famous.

Special offer includes first anniversary album in this magazine ad.

45EY2 Phonograph

The 45EY2 was really an updated 45EY series using the new rp-190 changer. This has proven to be the most popular model manufactured by RCA Victor. It was reasonably priced because there were no hinges or cover to deal with. The cabinet is simple yet good looking with rounded corners. RCA Victor did some cost cutting though. This model is not connectorized so repair is more difficult. Originally selling for $27.95, the model was so popular it was in production from 1950 through 1954 and then continued with a facelift in 1955 as model 6EY1.

A collage of '50s stuff in this Fender guitar ad includes the best selling 45EY2.

45EY2 Phonograph, RCA Victor's best selling model. Table model in dark brown Bakelite. Volume control located on right side. Originally priced at $27.95. $200-$300.

Dealer sheet for QEY4. This was a 45EY2 marketed for outside the USA. *Courtesy Lee Wells Collection.*

RCA VICTOR QEY4

El Sistema Moderno de Tocar Discos...

Fonógrafo "Victrola" Completo de 45*

Es indescriptible el placer que se experimenta al tocar discos con el nuevo Fonógrafo Victrola* QEY4. Es una magnífica audición de música grabada . . . sin que sea necesario prestar más atención al instrumento. En seguida se dará usted cuenta de las cualidades modernas y sobresalientes de este instrumento y de los maravillosos discos de "45 RPM". Ponga en el eje central de gran tamaño una colección de Discos RCA Victor de "45" . . . tóquelos todos con sólo empujar una vez la perilla respectiva. Estos discos tienen también otras ventajas . . . son muy económicos . . . son irrompibles . . . y duran diez veces más que los discos corrientes. El parlante muy potente de este instrumento y los Tubos RCA Victor de Tipo Preferido son otros elementos que contribuyen a su magnífico funcionamiento.

Su tamaño muy compacto . . . reducidísimo . . . es otra ventaja del QEY4. Se puede trasladar fácilmente de la sala a la cocina. Su estilo impartirá una nota de delicada elegancia a su hogar . . . sea cual fuere el sitio donde se instale este instrumento. Además, quedará encantado al enterarse de su precio ínfimo. Sí, amigo, el Fonógrafo "Victrola"* QEY4 reúne todos los méritos que usted quiere ver incorporados en un fonógrafo. Adquiera el QEY4 . . . y su entrada triunfal en el seno de su familia será un mensajero de felicidad . . . de admiración . . . y de alegría para todos.

EL FONOGRAFO "VICTROLA"* de 45

45EY3 Phonograph

The 45EY3 was the first table model Bakelite set that had a lid to close over the record changer. There were several advantages to this. Some people liked the Art Deco look of the cabinet and had no desire to stare at a record changer. Needle chatter (the noise that emanates from the stylus touching the record) is greatly reduced when the cover is closed, and dust could be kept off the records that were on the record changer. The new rp-190 record changer was used and this model was the runner-up in popularity to the 45EY2. It was priced at $34.95 and was so popular it was in production from 1950 to 1954 and continued with a facelift in 1955 as model 6EY2. This model is not connectorized and is one of the most difficult units to work on. RCA Victor was smart to put a handle on the back of the case so it could be called a portable record player. The handle is made of plastic and does not hold up very well. Finding one with an intact handle is quite a prize. Interestingly, advertising for this model usually shows the rp-168 changer inside. However, I have never come across one of these with that changer; it is always the rp-190. Perhaps the advertising was done before the decision was made to use the rp-190.

45EY3 Phonograph. Portable table model in dark brown Bakelite with lid. Runner-up in popularity to the 45EY2. Volume control located on top right inside lid. Originally priced at $34.95. $200-$300.

45EY3 Phonograph shown ready to be carried anywhere.

9Y510, 9Y511 Radio and Phonograph Combination

This was a very popular radio/phonograph that combined a decent-sounding AM radio with the rp-168 changer (9Y511) or rp-190 changer (9Y510). Three-stage tone control was achieved by turning a wafer switch to one of the three positions: low, medium, or high tone. The tone control was provided separately for the radio and phonograph, so the wafer switch had six positions! The radio is the typical 5-tube superheterodyne[252] used in most receivers manufactured in this time period. The tuning knobs are thin, easily broken, and are usually missing when one of these phonographs is uncovered in the attic or basement.

Note: 9Y511 cabinets cannot be interchanged with 9Y510 cabinets even though they look identical. On the 9Y511 the record changer is fitted into the Bakelite, while on the 9Y510 the record changer has its own motorboard.

Color advertisement featuring the 45J, 45EY, and 45EY3. Notice that the 45EY3 shows the rp-168 changer. This changer was never used in this model but some early ads do show it this way.

Everyone's going "45"

It's the modern thing to do! Good-by to the old "spindle-seeking," adjusting posts and clamps, one-at-a-time loading. America found there's a *better way*—with "45." Here's the first completely new system of recorded music in 50 years. The changer works from inside the spindle—it's foolproof and trouble-free. You can pick up a whole stack of records with 2 fingers and drop them on—*blindfolded.* Press a button and the show's on! "45" is all play and no work.

The modern way to play records

This "Victrola" 45 phonograph has its own super-sensitive speaker. World's finest, simplest, *surest* automatic changer. No posts to adjust—all "45" records are the same handy 7" size. They store in a bookshelf, no special cabinet needed. 45EY1. **$27.95.**

"Victrola" 45 Personal is the perfect traveling companion. Plugs into any 60-cycle AC electrical outlet. Smart, lightweight case—with handle for easy carrying. Plays a stack of records automatically—up to an hour of music! 45EY3. **$34.95.**

Prices shown are suggested list prices subject to change without notice. "Victrola"—T. M. Reg. U.S. Pat. Off.

Listen to "THE $64 QUESTION" every Sunday night at 10:00 p.m. New York time over your NBC station.

Everyone's giving "45"

You get bonus records, too! With every "45" instrument you get your choice of a brand-new RCA Victor "45" record release *every month for 6 months.* At no extra charge, you'll get a book of 6 record coupons from your dealer when you buy your "45." You also get a 6-months' subscription to Picture Record Review Magazine.

RCA VICTOR

Division of Radio Corporation of America

WORLD LEADER IN RADIO... FIRST IN RECORDED MUSIC... FIRST IN TELEVISION

9Y510 Radio and Phonograph Combination. The cabinet is dark brown Bakelite with a gold painted half grill in the front. Volume is on the top left side and tuning is on the top right side. Low-med-high tone control is on lower left side. Originally priced at $69.95. $225-$325.

Children's Phonograph

In 1950 only one cartoon player was available from RCA Victor, model 45EY15. The dark Bakelite case was painted white and decals added. A cautionary note is on bottom of the cabinet regarding children and lead poisoning. Disney characters were displayed, featuring Snow White and the Seven Dwarfs along with others. These units were only available for one year so they are rare.

9Y510 shown with closed lid.

45EY15 table model featuring Disney characters. Bakelite case has been painted white and decals added. Originally priced at $29.95. $200-$325.

This national ad shows that even a child can operate the 45 player. *Courtesy Camden County Historical Society.*

This holiday season advertisement notes that "Everyone's giving RCA Victor '45'."

Two page color holiday ad shows even Santa listens to the RCA Victor "45".

"45" POP ALBUMS ARE OUTSELLING 78-RPM BY 4 TO 1

During the last five weeks, orders for popular albums on 45-rpm have outnumbered those on "78" four to one, accountants report.

This seems to prove that new purchasers of the good-looking, bargain-rate "45" players are laying in a glorious collection of the colorful, extremely-practical "45" discs.

Selections of the new fans are very general and cover all pop sets released in 1949.

This article appeared in the *RCA Victor Distributors' Record Bulletin* of Jan. 23, 1950. *Courtesy Bill Pauluh Collection*.

Chapter Six

1951-1953: Filling Out the Product Line

In 1951 RCA Victor focused on adding a table model with improved fidelity. Use of a more powerful amplifier and bigger speaker provided a much fuller sound on the model 45EY4. A special phonograph was also made available to dealers and a fresh new children's model was introduced.

45EY4 table top Deluxe Record Player with lid closed.

45EY4 Phonograph

This was considered the top of the line record player featuring push-pull[253] amplifier with bass and treble control, large eight-inch speaker for better bass response and a pilot light built into the top of the Bakelite case. Unlike the cheaper models, this model has separate on/off control for both the record changer and the amplifier and is connectorized for easier servicing.

45EY4 table top Deluxe Record Player. Square cornered Bakelite case in dark brown has clear plastic frame that covers the gold covered front grill. All models with lid feature the Nipper logo on the inside of the lid. Volume and tone knobs are located on the lower right side. Retail price in 1951 was $49.95. $250-$350.

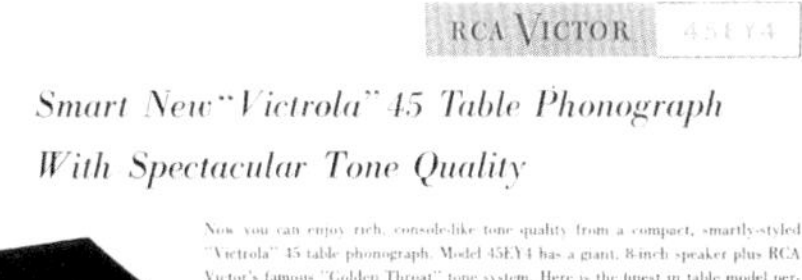

RCA VICTOR

Smart New "Victrola" 45 Table Phonograph

With Spectacular Tone Quality

Now you can enjoy rich, console-like tone quality from a compact, smartly-styled "Victrola" 45 table phonograph. Model 45EY4 has a giant, 8-inch speaker plus RCA Victor's famous "Golden Throat" tone system. Here is the finest in table model performance combining the distortion-free reproduction and ease of operation found only with RCA Victor's 45 system of recorded music. With only one hand, for example, you can easily load up to fourteen, 7-inch records on the convenient center spindle – for almost an hour of the music you want, when you want it.

Trimly-designed, the exciting RCA Victor 45EY4 will enhance your home . . . the gold-embellished plastic maroon cabinet blends beautifully with any decor. For those discriminating persons who desire impressive "45" performance in a smart table model, RCA Victor 45EY4 should be a unanimous choice.

"VICTROLA" 45 TABLE PHONOGRAPH

Dealer sheet for 45EY4. *Courtesy Lee Wells Collection.*

Advertisement trumpeting the success of the 45 system. *Courtesy Bill Pauluh Collection.*

15E Phonograph Demonstrator (Dealers Only)

Another addition to the line was a special high-performance table model 15E that was to be used by dealers only in their stores to play records for the customer. This unit has a nice push-pull amplifier driving a large twelve-inch speaker in the wooden cabinet. Sound quality was very good. There is a rumor that one of the questionable ways in which RCA manipulated its marketing of the 45 rpm record was to mismatch the impedance on the demonstration changer that played 78 rpm and 33 1/3, rpm records, thereby making the 45 record sound better! The 15E units are very hard to find today, since they were only available to dealers. The unit illustrated has a "see-through" red spindle cap on the 45 rpm changer. Not very many were equipped this way. Some stores had as many as six sound booths, and each one was equipped with a 15E Demonstrator. There were times when customers would take a stack of records into a sound booth and spend the afternoon listening without buying a thing! The record store that I frequented in Brooklyn, New York had one demonstrator behind the counter, and the salesman would play the record for the customer.

Pop show merchandiser from 1951.

CASH IN ON "45"

For replacements . . . use genuine RCA VICTOR crystals and styli

You can assure peak performance of RCA Victor 45 rpm's by using "original" RCA Victor crystals and styli.

Remember . . . RCA Victor originated the 45-rpm system—from stylus to record. And customers count on you to return their players to *original* high performance standards. By using *genuine* RCA Victor crystals and styli, you protect your servicing reputation and insure repeat business.

In addition to cartridges and styli for "45", there's a crystal cartridge and stylus specially designed for every model RCA Victor 78 or 78-33 rpm changer. Use them in your shop and avoid replacement problems.

. . .

Genuine RCA Victor crystals and styli are readily available from your local RCA Parts Distributor. SEE HIM TODAY.

New—Handy Stylus Dispenser and Utility Chest—Increase your stylus sales with this handsome, all-metal, 3-drawer chest—that doubles as counter dispenser and utility chest. Ask your **RCA Parts Distributor** today how you can get one, at no extra charge, with a model RCA Victor stylus inventory

TMK. ®

TT-8845

This advertisement appears in
Radio & Television Service Dealer—December, 1952

15E Phonograph Demonstrator shown in oak wood cabinet. This machine was only available to dealers. It is a large wooden table model that can play three speeds: 33, 45, and 78 rpm. Two record changers are mounted on the top and volume and tone controls are on lower right front. Well made push-pull amplifier drives a large twelve-inch front mounted speaker. Front speaker grill is adorned with large Nipper logo. Original price unknown. $1100-$1500. *Courtesy Mark Shoenthal Collection*

Radio & Television Service Dealer advertisement from December 1952.

45EY26 Children's Phonograph

A more modern children's player was also introduced, using the rp-190 changer. The model 45EY26, which featured Alice in Wonderland, was produced for several years and today is much more plentiful than the other cartoon character players that were manufactured for only one year. Bright colors were saved for the children, with colorful decals decorating the case and a bright red tonearm and turntable. All children's players used metal turntables instead of plastic. Special screws were also used, to prevent children from tampering with the insides of the players. The unit illustrated is a disguised 45EY2. The dark Bakelite case is painted white, the metal turntable is painted red, and the tonearm is red plastic.

RCA VICTOR

Tots and Toddlers Will Adore
The Complete "Victrola" 45 Phonograph

It's RCA Victor's new "Victrola" 45EY26! Decorated by Walt Disney in a gay Alice-In-Wonderland motif. Now the children may be entertained with their own special phonograph and records. RCA Victor's *Kiddies'* "45", the *complete* "Victrola" 45 Phonograph, brings a new, wonderful world of musical fairy tales and stories to all children. Here is continuous entertainment for all tots and toddlers! The RCA Victor *Kiddies'* "45" plays up to 14 non-breakable "45" plastic records with but one push of the starting button. Almost a full hour of continuous, trouble-free entertainment! Yes, even a small child can manipulate this fun-giving instrument.

The RCA Victor *Kiddies'* "45" has the famous RCA Victor "Golden Throat." The instrument is compact, light in weight. Take this fabulous RCA Victor phonograph home to the kiddies today!

THE KIDDIES' "45" PHONOGRAPH

Colorful dealer sheet featuring children enjoying model 45EY26. *Courtesy Lee Wells Collection.*

45EY26 Children's Record Player with Alice in Wonderland motif. Table model made of Bakelite that was painted white and decals added. Volume is controlled by the knob mounted on the lower right side. Original selling price $29.95. $225-$325.

Holiday advertisement for RCA Victor 45 players and multispeed players.

Chapter Seven

1954: New Radio Phonograph and Special Packages

In 1954 RCA Victor came out with a new table top radio/ phonograph model 4Y511 to replace the 9Y510. Instead of a slide rule dial, the new model has a circular dial mounted on the front. The case is made of black Bakelite. The only other change this year was an upgrade to the top of the line 45EY4. Its front grill was changed so that it resembled a television set. The robust four-tube push-pull amplifier with tone control was updated to use modern miniature vacuum tubes. A new ceramic pickup cartridge was also introduced with the upgraded model 45HY4. These updates and modifications yielded a much improved sound quality.

45HY4 table top Deluxe Record Player in Bakelite case. Front styling mimicked the look of a modern television set. Volume and tone controls are mounted on lower right side. This is considered one of the best sounding table top 45 rpm units that RCA Victor produced. Originally priced at $69.95. $250-$350.

4Y511 table model radio/phonograph in black Bakelite case. Under the front mounted dial there are three knobs for volume, tuning, and switching from radio to phono. The hinges are improved on this model and have a built in lid stop. Originally priced at $59.95. $250-$350.

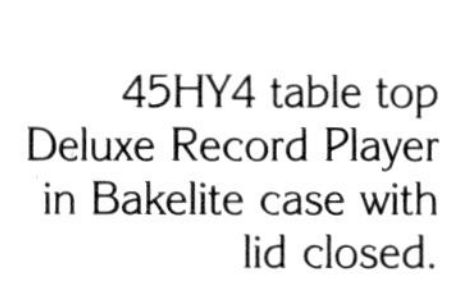

45HY4 table top Deluxe Record Player in Bakelite case with lid closed.

Advertising Packages

One of RCA Victor's techniques for bringing in new customers was to offer special prices on "package deals." Album sets such as *Treasure Chest of Mood Music* and *Listener's Digest* were sold with the popular model 45EY2. Once a customer had purchased the 45 rpm player, he or she was very likely to be a steady 45 rpm record customer. After all, the 45 changer played only 45 records!

Treasure Chest of Mood Music album.

Treasure Chest of Mood Music booklet.

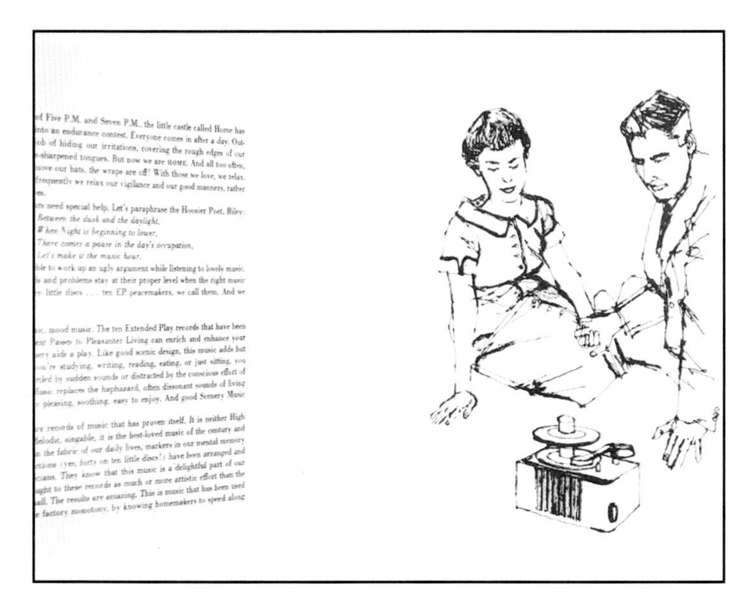

Sketches contained in the *Treasure Chest of Mood Music* booklet.

Classical music was featured in the *Listener's Digest* album.

National Geographic ad featuring special package that included the *Listener's Digest* album.

Centerfold ad for the holiday season.

Chapter Eight

1955: Product Line Completely Redone

In 1955 RCA Victor made a major shift in the production of its 45 rpm phonograph cabinets—from Bakelite or wood to a choice of Bakelite, wood, or plastic. This enabled the introduction in that year of a more varied and complete line of colorful models than had been possible with Bakelite only cabinets. The new models included battery-powered portable radio/phonographs and "Slide-O-Matic" units, which would accept the record through a slot in the front of the unit.

The corporation also introduced cost-cutting measures. Mute switches were removed from all models. The cycling cam rubber wheel was modified so that less rubber could be used in its manufacture. Anti-jamming refinements that had been added to the record changer mechanism in the previous four years were also removed to reduce costs.

The Skipper Model 6BY4

This unusual portable radio/phonograph plays one record at a time. The model 6BY4 was housed in RCA Victor's first plastic, rather than Bakelite, phonograph cabinet. The plastic was very soft and breakable, so most of the units found today are heavily scratched and may be cracked. This model was available in two color combinations: a subdued black and gray, and a flamboyant pink and white. A unique record storage compartment is located under the top lid and holds about a dozen records. A rubber puck mounted under the lid can be hand-tightened so that its expansion will hold the records firmly in place. The front grill shows off the famous Nipper logo, and volume and tuning knobs are located on the top front. The slide rule radio dial was unique in that the station frequency numbers were represented twice—once on top of the cabinet and again on the lid. This permitted the numbers to face the user when the lid was open or show the numbers upside down when the lid was closed, thereby allowing the user to read the numbers easily if the unit were sitting on its end. The phonograph is equipped with a new ceramic pickup cartridge, which improves the sound immensely. Battery power is provided by separate "A" and "B" batteries and four "D" cells, or an optional AC power supply that plugs into the wall and connects via a 7 pin connector on the rear of the cabinet. Most people did not buy the power supply, however, so they are very hard to find today. The power supply is about a foot long by three inches wide by two inches high—somewhat larger than today's commonly-used power supply units. A 6-volt DC motor was used to run the turntable. These motors did not maintain speed very well.

Model 6BY4A portable radio/single play phonograph called the "Skipper." Plastic cabinet in black and gray. Volume and tuning knobs are on opposite sides of slide rule dial residing on top front. Inside the phonograph area there are knobs for battery/AC operation and a phono/radio switch. Super heterodyne receiver drives front mounted four by six inch speaker. Optional AC power supply was available but hardly ever purchased. Original selling price $59.95. $125-$200.

Model 6BY4A portable radio/single play phonograph shown in upright position and ready for travel.

Model 6BY4B portable radio/single play phonograph called the "Skipper." Plastic cabinet in pink and white. This beauty is new right out of the box with all papers and batteries. Original selling price $59.95. $150-$225. *Courtesy Robert Becker Collection.*

Instruction booklet for 6BY4 series was unique in that there was a hole punched out in the center like a 45 record. The booklet was mounted on the record holder in the inside of the lid. *Courtesy Robert Becker Collection.*

Model 6BY4 series shown side by side.

Updating the Popular 45EY2 and 45EY3

The 45EY2 and 45EY3 were by far the best selling phonographs in the 45 rpm line from 1950 through 1954. Therefore, when the line was upgraded in 1955, a decision was made to maintain production of the two earlier models and just freshen up their looks. Besides using a different color scheme on the tonearm and spindle cap, a medallion was added to the tonearm. Most medallions show Nipper, but the international machines (designated with Q) have "RCA" on the medallion and a voltage switch inside the cabinet to select 110 volts or 220 volts. The gold motorboard color was lightened up and the Bakelite cabinet was changed from dark brown to black. Despite the greater durability of their Bakelite cabinets, these refurbished models did not sell well because of the availability of more colorful plastic models.

6EY1 and 6EY1Q Phonograph. Table model with black Bakelite cabinet. Volume control mounted on lower right side. This is an updated 45EY2. Original selling price $34.95. $225-$325.

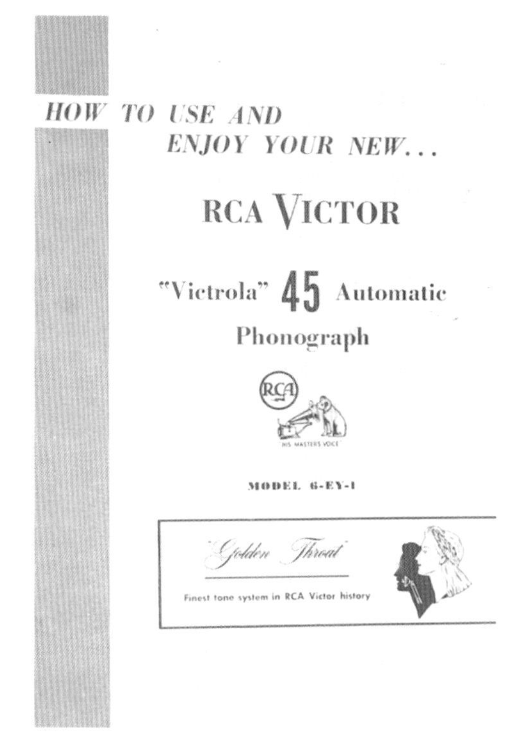

Instruction book for model 6EY1.

6EY2 Phonograph. Table model with black Bakelite cabinet and lid. Volume control on top front right. This is an updated 45EY3. Original selling price $49.95. $225-$325.

New Two-tone Portable Phonographs

A very popular and "smart-looking" design produced a vinyl covered portable record player available in three different two-tone color schemes: brown, green, and blue. The amplifier featured thumb wheels for volume and tone control, and two-stage audio provided decent volume to the four-inch speaker. The 6EY3A model featured brown and tan colors and was the most popular seller for the portables. The 6EY3B featured light and dark green-colored vinyl and was also a good seller. The 6EY3C in light and dark blue was only available in 1955 so they are hard to find today. For some reason RCA Victor decided to make the front of the 6EY3C the lighter color; the other two models in this series had the darker color on the front. The record changer was given an up-to-date look with a taupe-colored pickup arm and spindle cap.

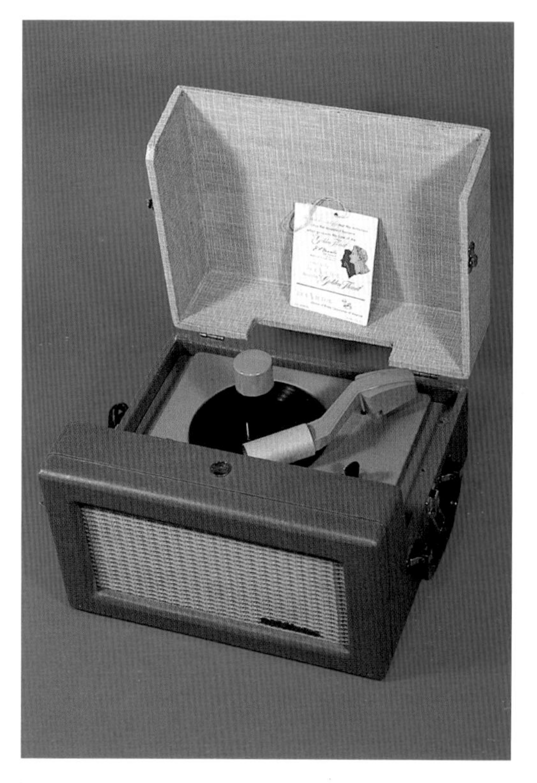

6EY3A portable record player. Vinyl covered portable in two-tone brown. Thumb wheels for volume and tone control are mounted in front of record changer. Original selling price $42.95. $150-$225. *Courtesy Dave Sica Collection.*

New old stock 6EY3A portable record player shown with original box. *Courtesy Dave Sica Collection.*

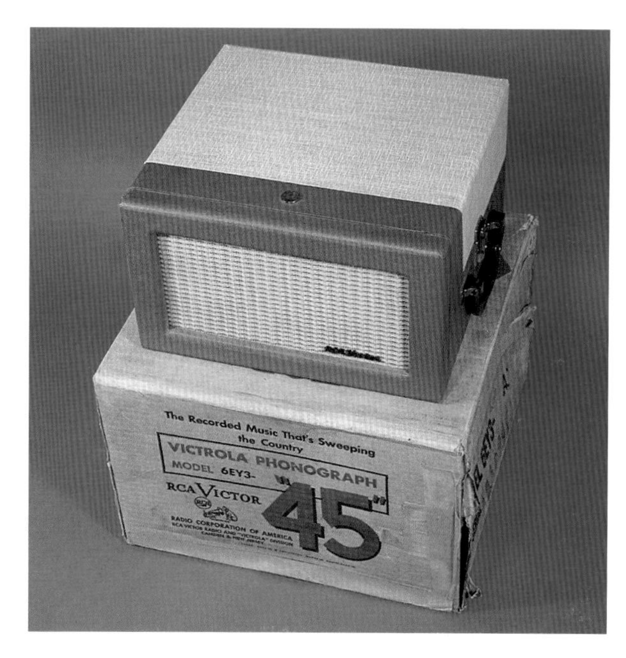

6EY3B portable record player. Vinyl covered portable in two-tone green. Thumb wheels for volume and tone control are mounted in front of record changer. Original selling price $42.95. $150-$225.

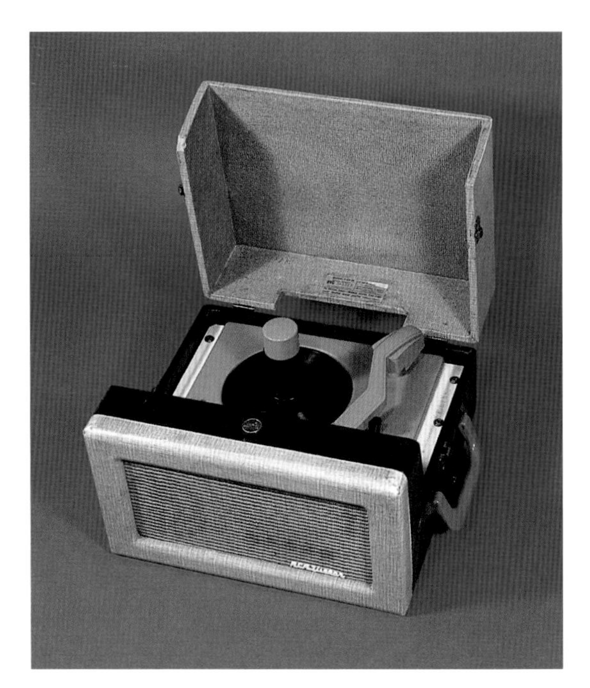

6EY3C portable record player. Vinyl covered portable in two-tone blue. Thumb wheels for volume and tone control are mounted in front of record changer. Original selling price $42.95. $175-$275.

Ding Dong School Phonograph

This model was based on the 6EY1 and featured decals of the popular TV show from the 1950s, *Ding Dong School*. The decals feature the teacher, Miss Francis, and a school bell. Not many of these phonos can be found today with the decals intact. Growing children could not resist the urge to deface the teacher's face and the school bell as they got older. The changer features a unique color scheme with an orange pickup arm and spindle cap, and a cream-colored motorboard. The arm also features the Nipper logo. Instead of plastic, the turntable was metal, which was typical for children's players.

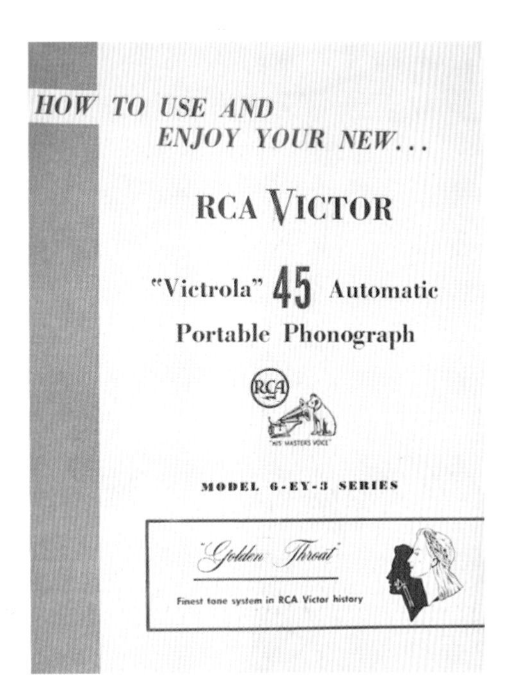

HOW TO USE AND ENJOY YOUR NEW...

RCA VICTOR

"Victrola" 45 Automatic Portable Phonograph

RCA

"HIS MASTERS VOICE"

MODEL 6-EY-3 SERIES

"Golden Throat"

Finest tone system in RCA Victor history

Instruction booklet for 6EY3 series.

6EY15 Ding Dong School phonograph. This table model was based on the 6EY1 and featured decals of the popular 1950s TV show, *Ding Dong School*. Volume control is on the lower right side of the black Bakelite cabinet. Original selling price unknown. $225-$325.

Grouping of RCA Victor children's players. From left: Ding Dong School (6EY15), Disney (9EY35), Roy Rogers (9EY36), and Alice in Wonderland (45EY26). In the center is a dealer display designed for children that holds 45 rpm records. *Courtesy Mark Shoenthal Collection.*

The "Slide-O-Matic"

In 1954 RCA Victor's Benjamin Carson completed design on a new single-play mechanism that would accept a 45 rpm record through a front slot. The user would then lift a metal lever and the record would start to play. When the record was finished, the lever would automatically return to the down position and the record would stop rotating. At this point it could be removed and another record inserted. If you wanted to remove the record while it was playing, you would push the lever down manually. Then you could remove the record.

People soon discovered that the novelty of sliding the record in and out of this fascinating new player was soon replaced by the awareness that it was a nuisance to constantly change the one-at-a-time records. This "aggravation" factor appears to have resulted in a low level of sales for the Slide-O-Matic. Despite its initial "gadget appeal," it did not become a universally popular unit. The case is made of soft plastic that scratches quite easily. Stress cracks usually occur where screws enter the cabinet. The Slide-O-Matic was available in many color combinations, as shown here. One special unit sported Ding Dong School decals.

Slide-O-Matics were also combined with AM radios. These table model units contained their own amplifiers and speakers, and were available in two color combinations. A five-tube superheterodyne receiver drives a four-inch speaker that is mounted on the side of the cabinet.

6JM25 Ding Dong School Slide-O-Matic attachment. White plastic cabinet with orange front and special decals. Center front latch would engage or disengage tonearm from record. Original selling price $14.95. $100-$175.

6JM1 Slide-O-Matic attachment. Black plastic case. Center front latch would engage or disengage tonearm from record. Original selling price $12.95. $75-$120.

6JM2 Slide-O-Matic attachment. Two-tone plastic cabinet, maroon case with contrasting beige front. Center front latch would engage or disengage tonearm from record. Original selling price $14.95. $75-$120.

6JM2 Slide-O-Matic attachment. Two-tone plastic cabinet, black case with contrasting maroon front. Center front latch would engage or disengage tonearm from record. Original selling price $14.95. $75-$120.

6JM2 Slide-O-Matic attachment. Two-tone plastic cabinet, off-white case with contrasting tan front. Center front latch would engage or disengage tonearm from record. Original selling price $14.95. $75-$120.

6JM2 See through Slide-O-Matic attachment. Usually only available to dealers. Very rare. $300-$400. *Courtesy Bob Havalack Collection. Photo by Tim Fabrizio.*

Burstein-Applebee flyer showing blowout price of $6.88 for the Slide-O-Matic.

Magazine advertisement showing the new Slide-O-Matics.

RCA VICTOR HAS A GIFT FOR MAKING PEOPLE HAPPY!

Give "The Gift That Keeps On Giving." Here are a few of many RCA Victor sets you'll find—priced just right—at your dealer's.

Playing Santa for someone special? Make sure your gift will please. Get an RCA Victor radio, "Victrola" phonograph or combination. It's one gift anyone will love at first sight—and for years to come.

Start your Christmas shopping on this page. Then have your dealer show you *all* the wonderful new RCA Victor table radios, clock-radios, portables, "Victrola" phonographs and record players.

From $14.95 to $35

Slide "45" record in slot and flip "Play-Bar." Red-and-white case with picture of Miss Frances of Ding Dong School.* Plugs into radio, TV phono-jack. (6JM25), $14.95
Other "Victrola" 45 players from $12.95.

SOOTHES YOU AWAKE. The *Roommate*, RCA Victor's lowest priced clock-radio. *Three* colors: black, antique white, turquoise green. (6C5), $29.95

TRAVELING COMPANION. Non-breakable "Impac" case! The *Skyway*, 3-way portable in two-tone gray, two-tone green or ivory. (6BX6), $34.95
Other portable radios from $24.95.

From $35 to $60

SMALL SIZE, BIG VOLUME. Radio-"Victrola" 45 Slide-O-Matic combination. Only 5⅜" high. "Golden Throat" tone. Black with gray or antique white with turquoise. (6XY5), $44.95

LOOK—NO ELECTRIC CORD. Now you can play your records anywhere—on *batteries*. It's the *Skipper*, RCA Victor's exciting new battery-operated "Victrola" 45 *phonograph and radio* combination. In sparkling white with bright red grille or velvety black with gray grille. (6BY4), $59.95

From $60 to $120

HI-FI THAT TRAVELS. Portable New Orthophonic High Fidelity "Victrola" 3-speed phonograph. Panoramic 3-speaker system. In rich brown leatherette. (7HFP1), $119.95

DOUBLE ENJOYMENT. Twin-speaker AM-FM radio—the *Lindsay*. Superb reception. "Golden Throat" tone. Phono-jack. Ebony finish. (6XF9), $69.95
Other table radios from $19.95.

PLAYS ALL THREE . . . 45, 33⅓ and 78 rpm records. Famous "Golden Throat" tone. Smartly styled cabinet in rich, black finish. (6ES3), $64.95
Other 3-speed phonographs from $29.95.

RCA VICTOR
RADIO CORPORATION OF AMERICA
CAMDEN 8, NEW JERSEY

*Ding Dong School is a National Broadcasting Company Inc. Service and trademark.

Manufacturer's nationally advertised prices shown, subject to change without notice. Slightly higher in far West and South. • See Milton Berle, Martha Raye alternately on NBC-TV, 2 out of every 3 Tuesdays. And don't miss NBC-TV's spectacular "Producers' Showcase" in RCA Compatible Color or black-and-white, Monday, December 12.

6XY5A Slide-O-Matic phonograph and radio. Two-tone plastic cabinet, black with contrasting gray front. Center front latch would engage or disengage tonearm from record. Volume and tuning controls are on each side of the front slot that holds the record. Original selling price $44.95. $200-$300

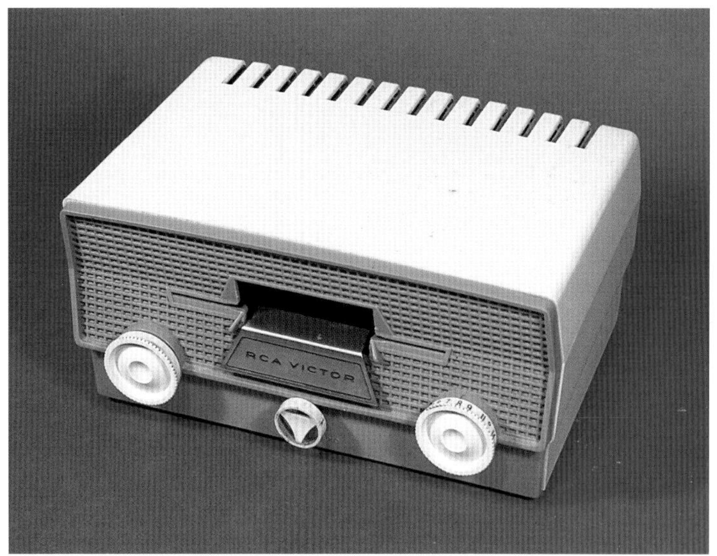

6XY5B Slide-O-Matic phonograph and radio. Two-tone plastic cabinet, white with contrasting aqua front. Center front latch would engage or disengage tonearm from record. Volume and tuning controls are on each side of the front slot that holds the record. Original selling price $44.95. $200-$300.

6XY5A and B Slide-O-Matic phonograph with radio.

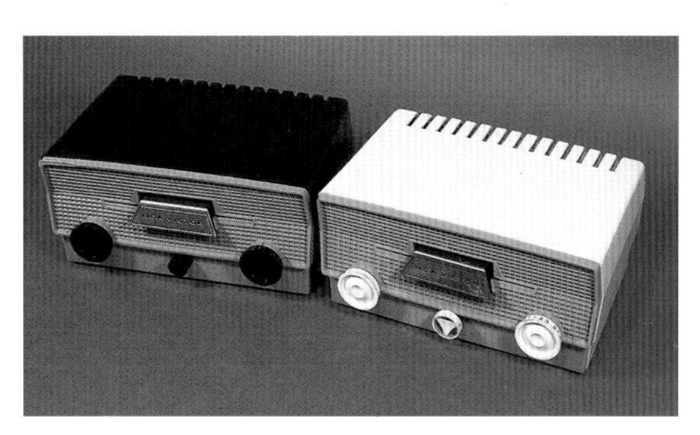

Restyled 45 RPM Attachments

The 45J2 attachment had been available from 1950 through 1954 and was selling in any color as long as it was dark brown Bakelite with a black tonearm and red spindle cap. For 1955 RCA Victor introduced the 6JY1 series, a face-lifted version that was available in three different color combinations. Tonearms now matched the spindle cap in ivory. The Nipper logo was prominently displayed on every tonearm, and the medium gold motorboard was now a light gold. A new Bakelite cabinet featured flared handles on the sides so it was easier to carry the unit.

6JY1A attachment with black Bakelite cabinet. Changer features ivory tonearm and spindle cap. The Nipper logo was also prominently displayed on the tonearm. Original selling price $19.95. $75-$120.

6JY1B attachment with white painted Bakelite cabinet. Changer features ivory tonearm and spindle cap. The Nipper logo was also prominently displayed on the tonearm. Original selling price $19.95. $75-$120.

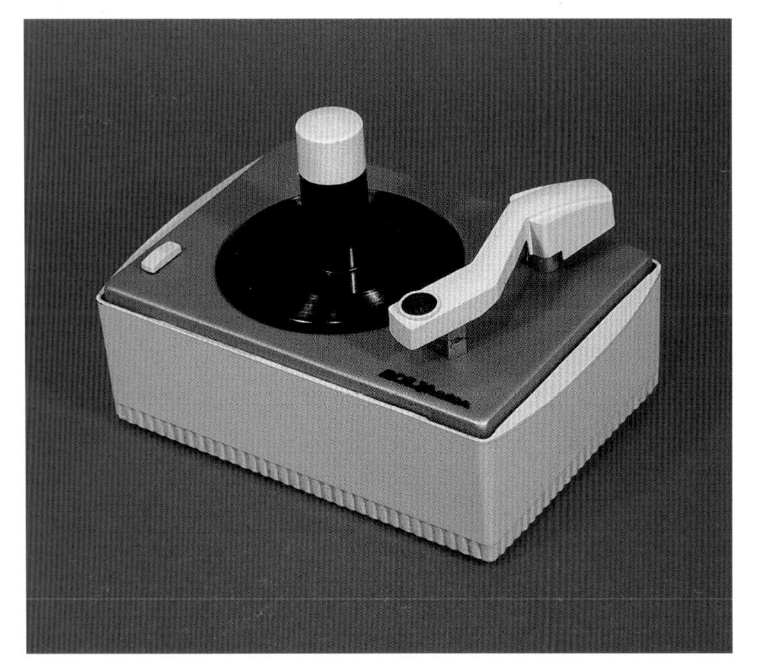

6JY1C attachment with green painted Bakelite cabinet. Changer features ivory tonearm and spindle cap. The Nipper logo was also prominently displayed on the tonearm. Original selling price $19.95. $75-$120.

For some reason RCA Victor offered another attachment with a gray tonearm and spindle cap and enclosed in a black Bakelite case. International attachment model numbers were followed by the letter "Q". These did not display Nipper, but instead the letters "RCA" on the tonearm. These units included a sleeve that would change the speed of the motor depending on whether it was using a frequency of 50 or 60 hertz (cycles per second). A voltage switch was also mounted underneath the motorboard for 110 volt or 220 volt operation.

6JY1 and 6JY1(Q) attachment featuring International Option and a black Bakelite case. International models would be followed by a "Q". These would not display Nipper but instead the letters "RCA" on the tonearm. Original selling price $19.95. $75-$120.

Grouping of 6JY1 series attachments shown with Nipper.

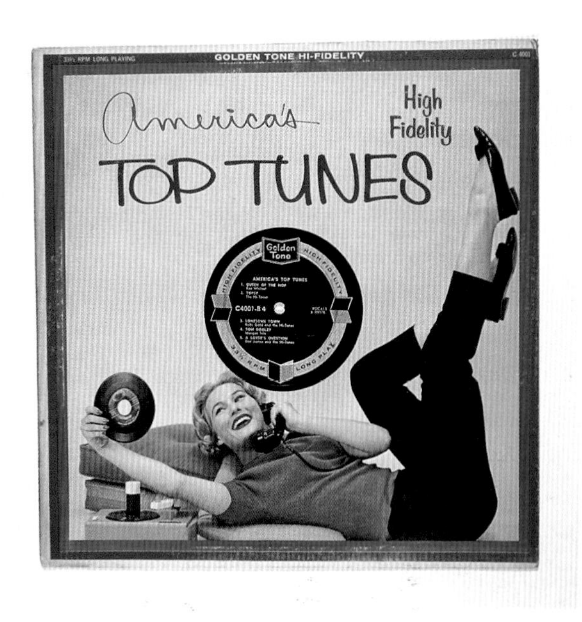

Album cover from late 1950s featuring model 6JY1C.

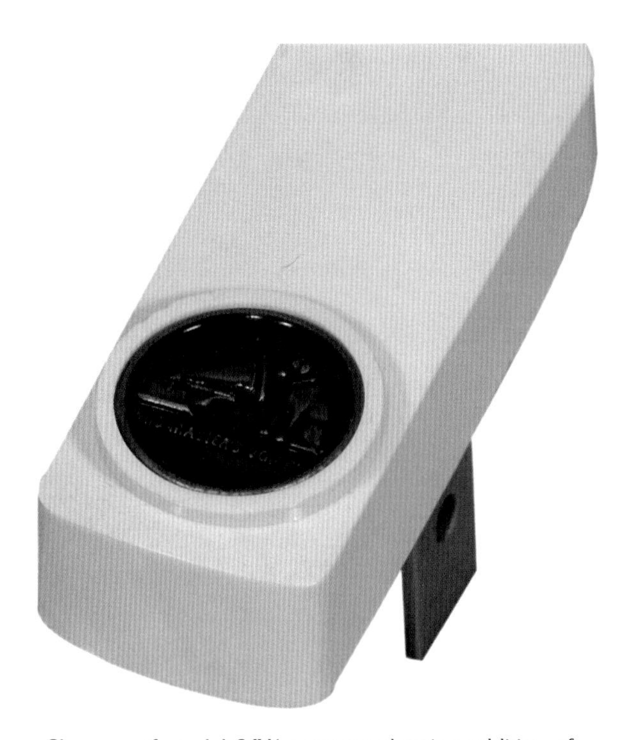

Close-up of model 6JY1 tonearm showing addition of Nipper medallion.

Colorful 7EY Series

These very popular models featured "gay pastel colors" and were available in regular and deluxe variations. The cabinet was made of soft plastic and easily scratched. The deluxe models had more powerful amplifiers, larger speakers, and a peach-colored tonearm and spindle cap. Over time, the tonearm tended to fade to an orange color, making the color combination look hideous. Regular models had ivory-colored tonearms and spindle caps. Teenagers loved these machines because they looked so cool. Sound quality was downright tinny, even with the deluxe upgrades. An optional vinyl-covered hard or soft carrying case was available at extra cost, to safely carry your 7EY series record player.

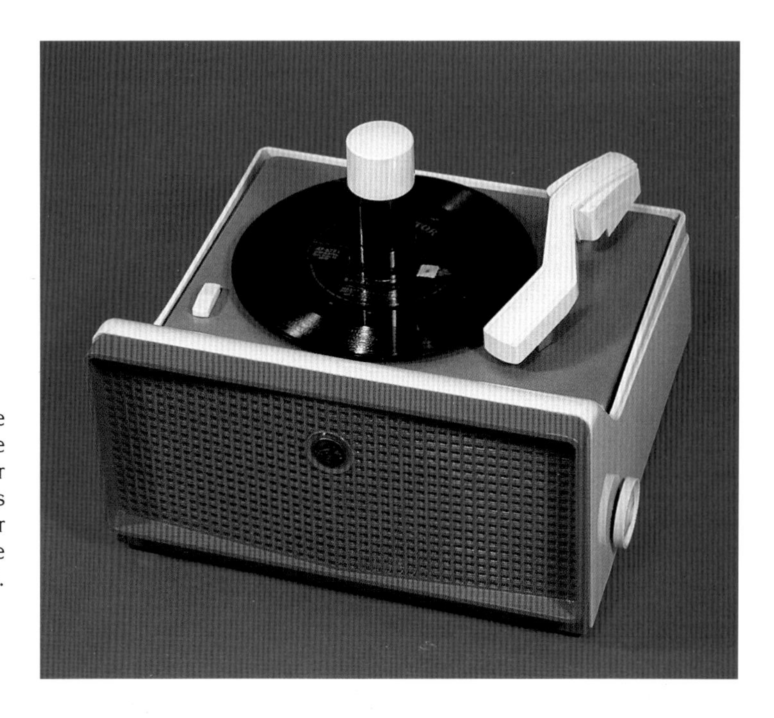

7EY1EF table model record player. White plastic cabinet with flame red grill. With age the red looks more like pink. Very popular model with the teenage set. Volume control is located on lower right side. One tube amplifier drives four-inch speaker. Original selling price $32.95. $200-$300.

7EY1JF table model record player. Gray plastic cabinet with coral grill. Volume control is located on lower right side. One tube amplifier drives four-inch speaker. Original selling price. $32.95. $200-$300.

7EY1DJ table model record player. Black plastic cabinet with gray grill. Volume control is located on lower right side. One tube amplifier drives four-inch speaker. Original selling price $32.95. $200-$300.

Color magazine insert featuring budget priced Victrolas.

7EY2JJ Deluxe table model record player with two audio stages of amplification and a larger four by six inch speaker. Dark gray plastic cabinet with light gray front grill. Volume control is located on lower right side. Record changer color scheme is quite vivid with peach-colored spindle cap and tonearm. Original selling price $36.95. $200-$300.

7EY2HH Deluxe table model record player with two audio stages of amplification and a larger four by six inch speaker. Dark green plastic cabinet with light green front grill. Volume control is located on lower right side. Another vivid color scheme with peach-colored spindle cap and tonearm. Original selling price $36.95. $200-$300

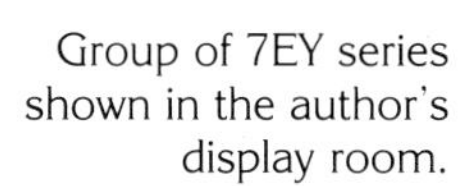

Group of 7EY series shown in the author's display room.

Chapter Nine

1956: New Orthophonic High Fidelity Models Introduced

For 1956 RCA Victor improved the sound of their 45 rpm phonographs by introducing New Orthophonic High Fidelity. They produced the first unit considered to be true high fidelity, proudly displaying the label on the front grill. The new high fidelity was accomplished by using a four-tube amplifier with two audio stages and push-pull output, a Sonotone ceramic pickup cartridge, and separate woofer and tweeter. Miniature vacuum tubes were used and the amplifier chassis was quite compact. In this case there was plenty of room inside the cabinet so the small chassis does nothing but make servicing more difficult. The best frequency response is heard when the top lid is closed. The amplifier also featured bass and treble control. A record storage area was provided on the left side for about twenty-five records. The cabinet was available in three different woods: mahogany, maple, and oak. By far the most popular color was mahogany. It is frequently difficult to distinguish the maple cabinets from the oak models. However, I have found that the oak model always has light colored rubber feet and the inside of the lid appears to have been painted.

A new Orthophonic High Fidelity portable model was also available this year. The vinyl case was brown and tan with a wraparound grill. All the same components are present in this model, which had the added convenience of portability.

7HF45 New Orthophonic High Fidelity table model phonograph in mahogany wood. Controls are located on the lower right side of the wooden cabinet. A record storage area is provided on the left side for about twenty-five records. Original selling price $79.95. $225-$325.

7HF45 New Orthophonic High Fidelity table model phonograph in mahogany wood with lid closed.

7HF45 New Orthophonic High Fidelity table model phonograph in maple wood. Controls are located on the lower right side of the wooden cabinet. A record storage area is provided on the left side for about twenty-five records. Original selling price $84.95. $275-$375.

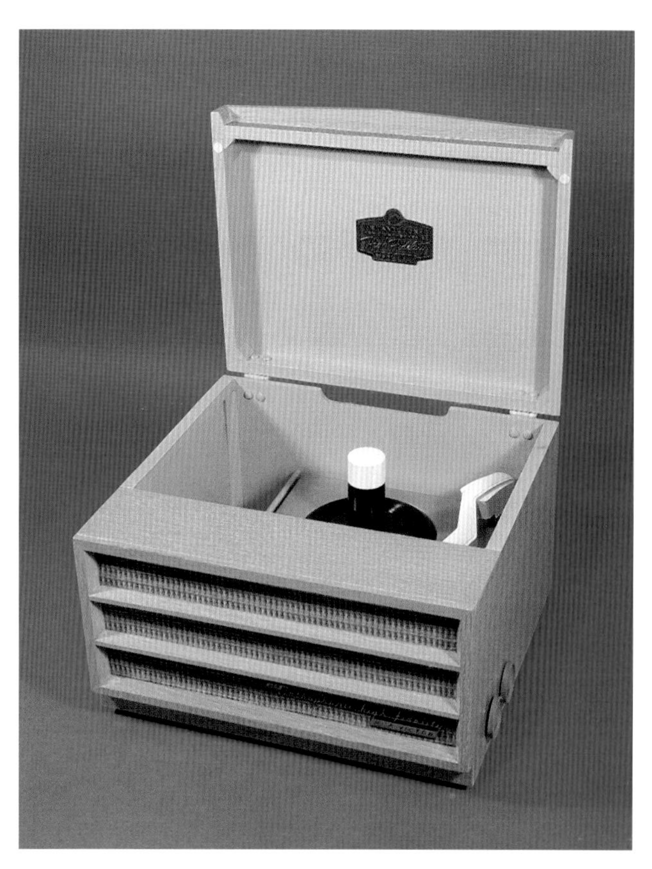

7HF45 New Orthophonic High Fidelity table model phonograph in oak wood. Controls are located on the lower right side of the wooden cabinet. A record storage area is provided on the left side for about twenty-five records. Original selling price $84.95. $250-$350.

7HF45 New Orthophonic High Fidelity table model phonograph in maple wood with lid closed.

7HF45 New Orthophonic High Fidelity table model phonograph in oak wood with lid closed.

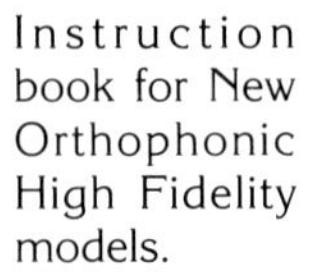

Instruction book for New Orthophonic High Fidelity models.

8HF45P New Orthophonic High Fidelity portable phonograph. Vinyl covered portable model containing the same insides as the 7HF45 including push/pull amplifier, dual speaker woofer and tweeter, etc., with the added convenience of portability. Wraparound grill gives nice effect. Tone and volume knobs are on the lower right side. Original selling price $69.95. $175-$275.

Magazine advertisement showing RCA Victor's product line.

16 great ways to sound your Christmas greetings!
New Sensations in Sound by RCA Victor—from $19.95

The *Kerry*. Table radio with printed circuits for longer life. Black, antique white, spruce green, pink. (6X5) $19.95

The *Newcliffe*. Twin-speaker "Golden Throat" tone. Phono-jack. Black, gray, turquoise, or ivory cabinet. (8X8) $29.95

The *Shipmate*. 3-way portable in non-breakable "IMPAC" case. Flame red, 2-tone green, 2-tone gray. (7BX5) $29.95*

The *Roommate* clock-radio. Wakes you with rich "Golden Throat" tone. Black, white, turquoise or pink. (6C5) $29.95

Portable 4-speed "Victrola."® Rich luggage styling in tan, tan-and-brown or blue. Plays on AC. (7EMP2) $29.95

The *Winsome*. Transistor portable with "Golden Throat" tone. "IMPAC" case in two 2-tone finishes. (8BT7) $39.95*

The *Merriweather* clock-radio. It swivels! Turns itself on, off. Runs appliances. 3 finishes. (8C8) $42.95

Deluxe automatic 45 "Victrola." Plays almost 2 hours of "45" EP's. "Golden Throat" tone. 2 finishes. (8EY4) $49.95

New "*Globe Trotter*" 3-way portable. Rotating "Wavefinder" antenna. "IMPAC" case in gray or aqua. (7BX8) $49.95*

Automatic 4-speed portable "Victrola." Two speakers. 2-tone gray or 2-tone green. AC operation. (7ES6) $79.95

The *Mark VII*. 4-speed Hi-Fi. Panoramic 3-speaker system. Stereophonic. Brown. (8HFP1) $129.95

The *Judicial*. Push-button Hi-Fi tape recorder. 3 speakers. 2 speeds. Voice-music switch. Gray case. (7TR3) $199.95

A $24.95 value for just $5.00! 60 All-time hits from
GLENN MILLER Limited Edition No. 1
when you buy one of these Fabulous 45 "Victrolas"!

Only the Fabulous "45" system offers so much—
EASIEST TO PLAY
MOST TROUBLE-FREE
PLAYS ALMOST 2 HOURS OF MUSIC WITH ONE LOADING OF "45" EP RECORDS

Lowest priced *true* Hi-Fi! The *Mark VIII*. Multiple speakers. 3 popular wood finishes. (7HF45) $79.95†

Automatic 45 "Victrola." 3 finishes. (7EY1), $32.95. Extra power, larger speaker, 2 finishes. (7EY2) $36.95

Automatic Portable "45" with tone control. Brown-and-tan or 2-tone green carrying case. (6EY3) $42.95

Give "The Gift That Keeps On Giving"

RCA VICTOR
RADIO CORPORATION OF AMERICA
CAMDEN 8, NEW JERSEY

YOU'LL HEAR SONGS LIKE: "String of Pearls" "Little Brown Jug" "Moonlight Serenade" "Lamplighter's Serenade"

*Less batteries. Always insist on RCA batteries—they're Radio-Engineered for extra listening hours. †Price quoted for Mark VIII for mahogany finish.
Manufacturer's nationally advertised list prices shown, subject to change. Slightly higher for West and South. Most models available in Canada.

Elvis Presley Model

A special version of the 6EY3 series portable phonograph was available as model 7EP45. Treatment included a different grill, color scheme, and Elvis Presley's signature in gold on the top front right corner. A special extended play (EP) record set was also part of the package and was not available in stores. Because this model was adorned with Elvis's signature, it is worth quite a bit more in today's collector market than other portable models.

7EP45 "Elvis" Portable Phonograph. Blue vinyl covered portable with Elvis's signature in embossed gold on top right. Amplifier has two audio stages and four-inch front firing speaker. Volume and tone knobs are located in front of record changer. Originally priced at $47.95. $450-$650. *Courtesy Dan Saporito Collection*.

Close-up of Elvis's signature. *Courtesy Dan Saporito Collection*.

Updated Deluxe Phonograph

The model 8EY4 was introduced as a replacement for the top of the line 45HY4 phonograph. A cost reduction was achieved by plugging the pilot light hole on the top of the Bakelite case and simply not providing the pilot light. The same robust four-tube push-pull amplifier with tone control was provided, with large eight-inch speaker for excellent tone quality. The record changer was given a facelift with an ivory-colored tonearm and spindle cap. The Bakelite cabinet was capped with a gray (8EY4DJ) or beige (8EY4FK) plastic front with Nipper logo in the center.

8EY4DJ Deluxe Phonograph. Brown Bakelite table model with gray plastic front. Same robust four tube push-pull amplifier that was used in 45HY4 with large magnet eight-inch speaker for excellent tone quality. Volume and tone controls on lower right side. Original selling price $49.95. $225-$325.

8EY4FK Deluxe Phonograph. Maroon Bakelite table model with beige plastic front. Same robust four tube push-pull amplifier that was used in 45HY4 with large magnet eight-inch speaker for excellent tone quality. Volume and tone controls on lower right side. Original selling price $49.95. $225-$325.

8EY4DJ Deluxe Phonograph with lid closed.

8EY4FK Deluxe Phonograph with lid closed.

Chapter Ten

1957: The Last Hurrah!

For 1957 RCA Victor updated their portable phonographs with new styling and a spectrum of cost reductions, to the extent that the players can be regarded as downright cheap! The vinyl used on the inside of the lid is so thin and insubstantial that it resembles paper rather than vinyl and does not hold up well. Only one tube of amplification is directed into a four-inch speaker, with the result that the sound quality is minimal. A high-output crystal pickup cartridge was used, which provided adequate volume.

An interesting table model television and 45 phonograph was available also that was just twenty-four inches wide. Called the "Bellevue," it was available in dark and light woods.

The Bellevue, model 14VT8155 wooden table model featuring twelve-inch black and white television, and 45 rpm phonograph. Shown in mahogany. *Courtesy Joe Centanni Collection.* $300-$400.

8EY31 Series Portable Phonograph. Vinyl covered cabinets were available in two color schemes, brown and tan (KE), or green and beige (HE). Volume and tone controls are provided on the lower right front. Original selling price $39.95. $125-$200.

Color magazine advertisement featuring Perry Como and a host of fine phonographs.

The End of an Era

It had been RCA Victor's dream not only to provide the public with a new record and record-playing system, but to create a system that would be a complete, long-term replacement for the 78 rpm record. The 45 rpm system was wildly popular for nearly a decade, and particularly so with teenagers, who found it convenient and cool to carry their portable players and stacks of seven-inch records to parties. Ultimately, however, the dream was unrealized, largely because of events discussed here. Customers finally were faced with choosing between RCA's 45 player, which played only one size and one speed of record, and a multispeed player that could handle all record sizes and speeds. Market share was shrinking for the 45-only machines. Furthermore, the new stereophonic disc was about to be introduced, and manufacturers were preparing machines to accommodate the new records. In the midst of all this a fire destroyed the factory in Chicago that was manufacturing the 45 rpm changers for RCA Victor. This must have been the final blow because new stereophonic 45 player designs were never made available to the public by RCA Victor. Voice of Music made a stereo 45-only machine, and licensed the changer out to Arvin and other companies. Some 45 rpm stereo records were produced, but they did not catch on. It is interesting that stereo 45s from this time period (late 1950s) never became popular, because the new multispeed stereo players could play them, along with records of other sizes and speeds.

Chapter Eleven

Licensing Helps Spread the New System

RCA Victor allowed other manufacturers to license and use the 45 rpm record changer. What better way to increase the sales of 45 rpm records, since the players would accommodate only this one-size, one-speed record? Many manufacturers jumped at the opportunity. In fact, I continue to find machines not previously encountered, so the list of companies keeps growing. It includes Birch, Capitol, Crescent, Decca, Emerson, Symphonic, Motorola, Zenith, Columbia, Teletone, Truetone, Steelman, Magnavox, Majorette, Phono Art, Harleigh, Voice of Music, Hudson, Phonola, Silvertone, Victory, and Westinghouse. These companies usually designed their own cases and amplifiers, and then placed the RCA Victor record changer inside their cabinets. A few companies used the same cabinets and amplifiers that RCA Victor used, and changed only the nameplate. Many of the companies used a more brightly-colored spindle cap than RCA, and one company (Victory) changed the color of the plastic tonearm to maroon.

Many of these machines do not have model numbers displayed. Wherever possible, I have included the model number in the caption.

Birch model unknown. Equipped with rp-190 changer. Green and white portable with tone control. Powered by a single stage amplifier, tone and volume controls are front mounted in center of silver grill. $100-$175.

Birch model unknown. Equipped with rp-190 changer. Powered by a single stage amplifier, the controls sit on the right side of a smart looking black grill. $100-$175

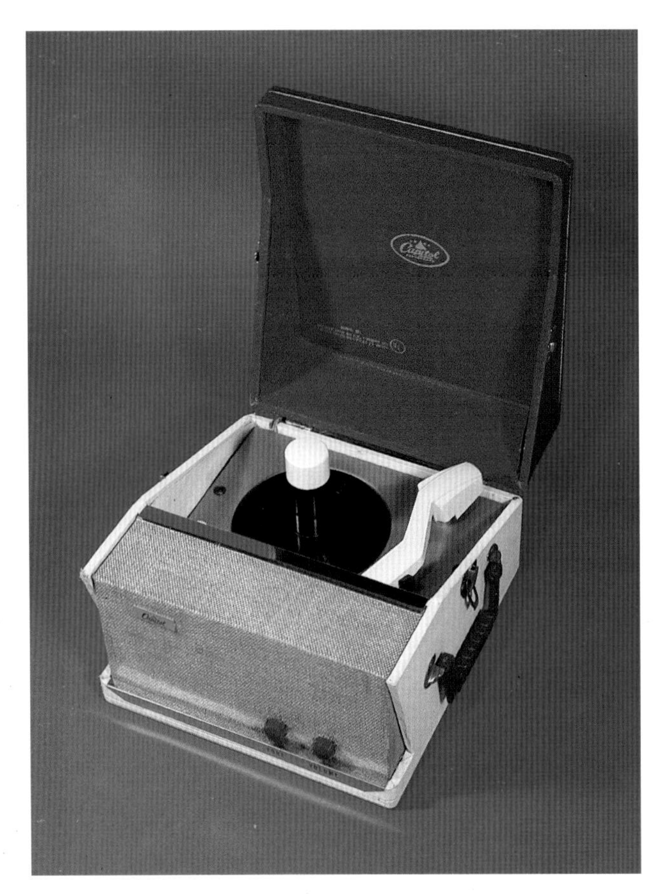

Capitol model 821. Portable model in red and white with rp-190 changer and tone control. Single stage amplification, white knobs. Amplifier section in front can be pulled out as a separate box for easy servicing. $100-$175.

Capitol models 821 and 721 together with lids closed.

Capitol model C45. Bakelite attachment with rp-168 changer. Identical to the RCA Victor attachment except for Capitol logo on the front left. On/off volume knob on right side of cabinet. $100-$150.

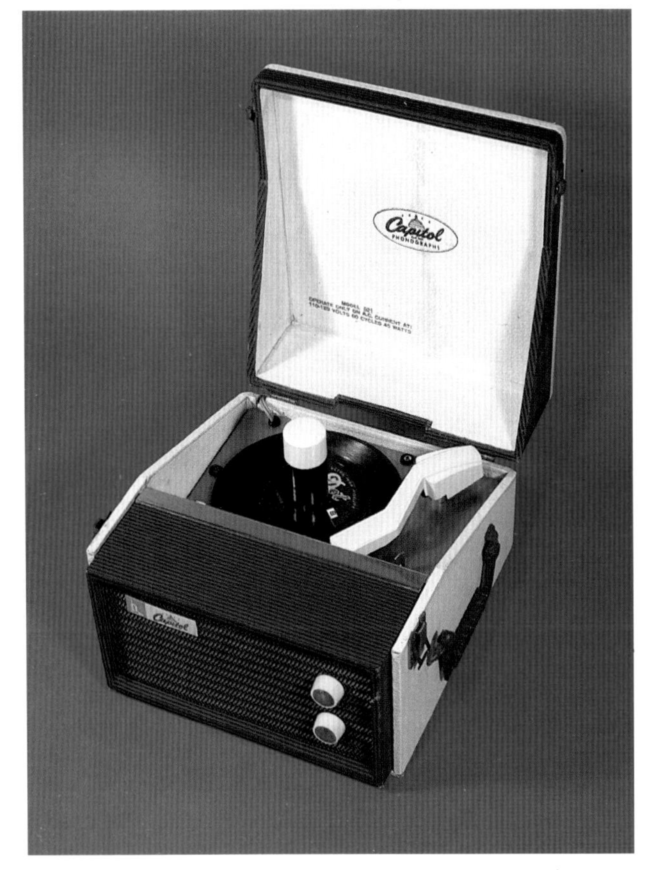

Capitol model 721. Portable model in red and white with angled sides at ends of front grill. Equipped with rp-190 changer. Single stage amplification, red knobs on lower right of silver front grill. $100-$175.

Chesapeake Bird Call model unknown. Black wooden portable cabinet equipped with rp-190 changer. Designed to be plugged into pre-1956 car's cigarette lighter (6 volts). Husky amplifier provides plenty of air filling power. Speaker is hooked up externally. Contains 6 volt DC converter to 120 volts AC. $250-$350.

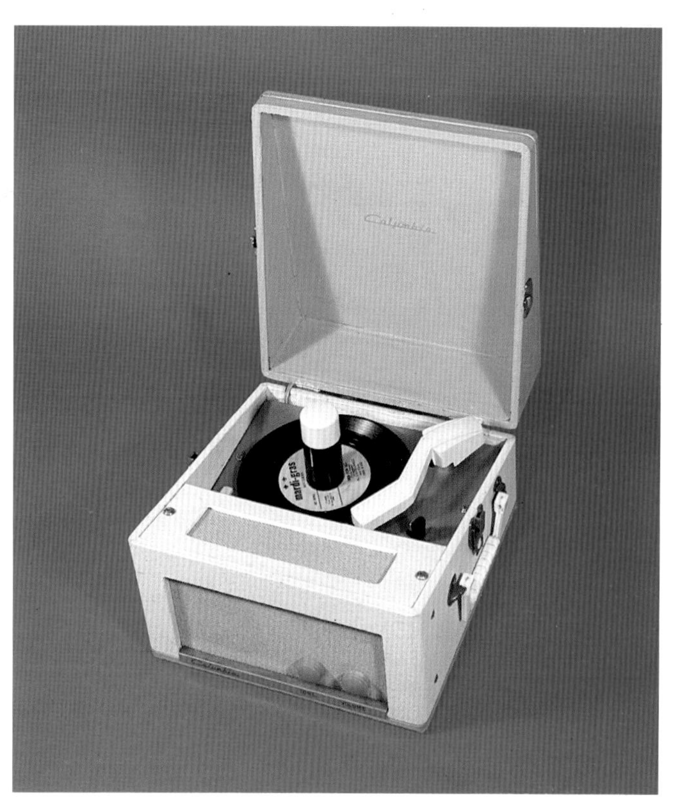

Columbia model 540. Green and white portable with rp-190 changer and tone control. Single stage amplification, green knobs on right front of light green grill. $100-$175.

Chesapeake Bird Call model with cover closed.

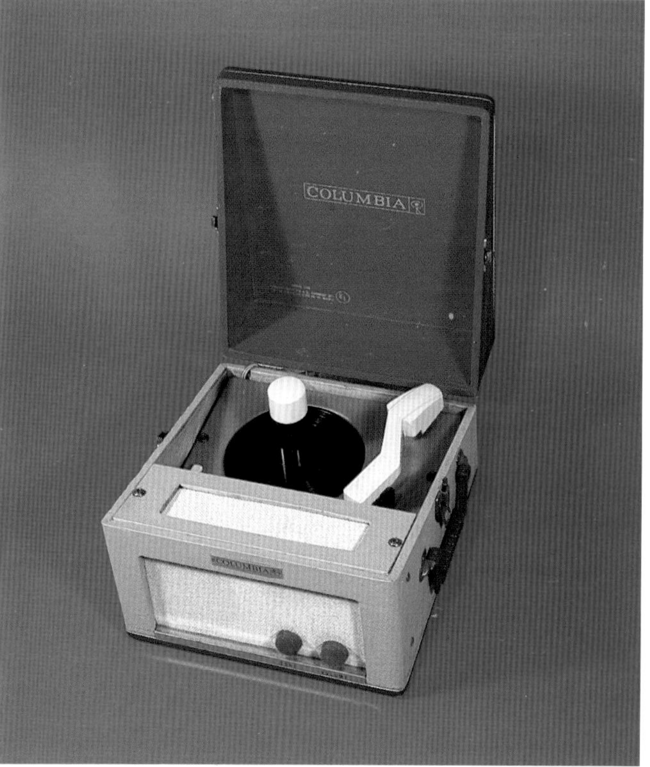

Columbia model 540. Red and white portable with rp-190 changer and tone control. Single stage amplification, red knobs on right front of white grill. $100-$175.

Concord "New World" model. Miniature wooden cabinet housing an rp-190 changer on top and a radio with slide rule dial near the bottom. Radio is accessible through lower front doors. Radio has four knobs under the dial. $175-$250.

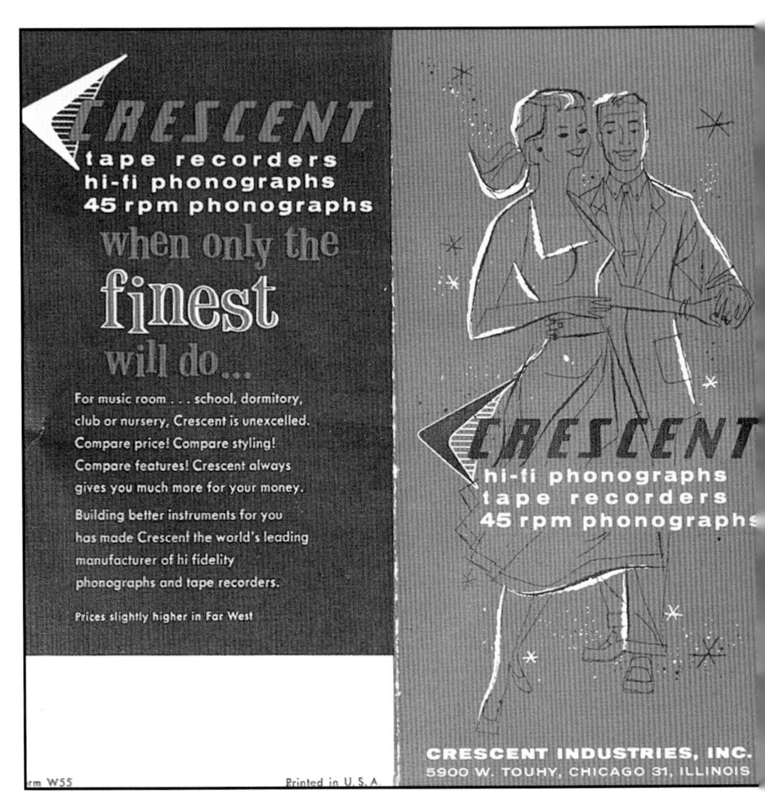

Crescent brochure from early 1950s. *Courtesy Ray Tyner Collection.*

Crescent model F737. Two-tone green table top unit with cream-colored knobs on top right for volume and tone. Equipped with rp-190 changer. Bottom of cabinet angles upward toward the front and is supported by chrome bar. Single stage amplification feeds bottom mounted speaker. $150-$225.

Concord "New World" model with front wooden doors opened showing radio.

Crescent with cartoons, model unknown. Model is identical to RCA Victor model 45EY26 except for logo and brighter spindle cap. Numbers on amplifier, cabinet, and bottom are all Crescent manufacture numbers. Equipped with rp-190 changer and two stage amplifier. Volume control mounted on lower right side. $250-$350.

Crescent brochure showing model F637. Notice where the volume control is. All examples of this machine I have come across have the volume control on the top, not the front.

Crescent model F637. Tan and tweed table top unit with cream-colored volume knob located on top front. Equipped with rp-190 changer. Bottom of cabinet angles upward toward the front and is supported by two chrome legs. Single stage amplification feeds bottom mounted speaker. $150-$225.

Crescent model unknown. Light green and tan table top unit with white volume and tone knobs on lower right side. Equipped with rp-190 changer. Two stage amplification drives front mounted speaker mounted in silver grill. $100-$175.

Crescent model 453A. Two-tone green table top unit with white knobs on right side for volume and tone. Equipped with rp-190 changer. Two stage amplifier with front silver grill. Some units had top cover with handle for portability. This model came with a lid equipped with a handle, making it portable. $120-$200.

Crescent model unknown. Brown and tan table top unit with brown volume knob on lower right side. Equipped with rp-190 changer. Two stage amplification, gold grill with gold frame. $100-$175.

Crescent model 453A with lid attached.

Crescent model unknown. Rp-190 attachment with two-tone metal cabinet black and gray hammertone. Logo mounted on left center of front panel. $120-$170.

Crescent model unknown. Rp-190 attachment with two-tone metal cabinet dark and light gray. Logo on top lower right. $100-$150.

Decca model DP910 with closed lid.

Decca model DP910. Portable red and black with tone control. Unique barrel shaped cabinet with white grill. Equipped with rp-190 changer. Single stage amplification, red knobs on lower right of front grill. $200-$300.

Decca model DP908. Table top model with gray paisley vinyl. Brown volume control is on lower right front. Equipped with rp-190 changer. Single stage amplifier with off-white front mounted grill. $100-$175.

Decca model DP911. Brown and white portable with tone control. Equipped with rp-190 changer. Single stage amplification, brown knobs on lower right two-tone grill. Lid features brown and white stripes. $100-$175.

Decca model P-903. Exact clone of RCA Victor 45EY3 with Decca name plate. $200-$300.

Decca model P-907. Exact clone of RCA Victor 45EY2 with Decca name plate. $200-$300.

Crosley model unknown. Bakelite attachment in dark brown with scalloping. Equipped with rp-168 changer. $150-$175.

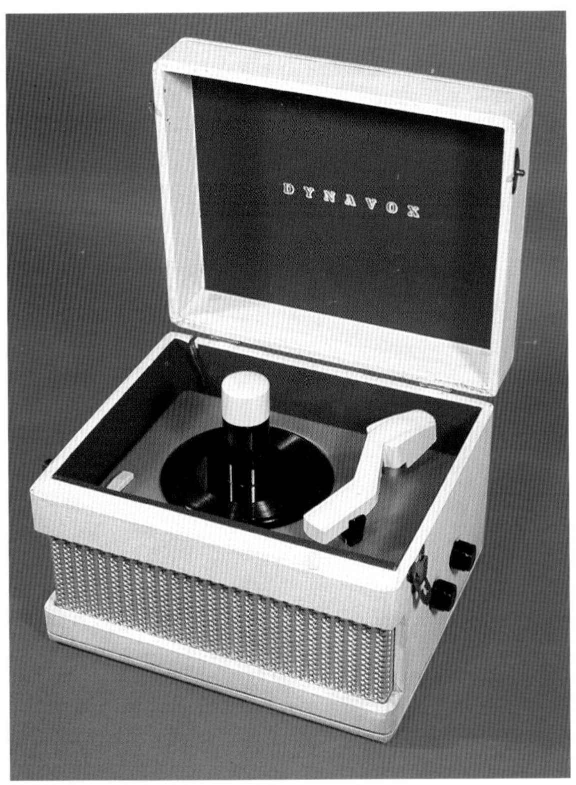

Dynavox model unknown. Portable red and cream with volume and tone control. Equipped with rp-190 changer. Single stage amplifier driving front mounted speaker and chrome grill. Brown knobs mounted on lower right hand side. Inside of lid is vivid dark red with Dynavox logo. $200-$300.

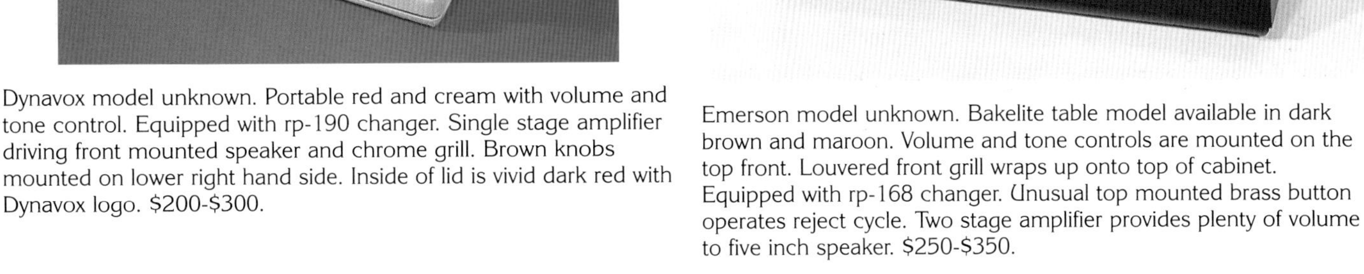

Emerson model unknown. Bakelite table model available in dark brown and maroon. Volume and tone controls are mounted on the top front. Louvered front grill wraps up onto top of cabinet. Equipped with rp-168 changer. Unusual top mounted brass button operates reject cycle. Two stage amplifier provides plenty of volume to five inch speaker. $250-$350.

Electronic model unknown. Portable brown and tweed with white volume control mounted on lower right front of white grill. Equipped with rp-190 changer. Very compact design with single stage amplifier. $100-$175.

Emerson model 623A. Wooden mahogany table model. Equipped with rp-168 changer. Volume and tone controls are mounted on the top front. Louvered grill wraps completely around cabinet. Two stage amplifier provides plenty of volume. $200-$275.

Emerson model unknown. Bakelite dark brown table model featuring radio and phonograph. Equipped with rp-168 changer. Volume and tone controls straddle slide rule radio dial on top front of cabinet. Reject button is mounted on right side. Five tube super heterodyne with single stage audio amplifier. $250-$350.

Color advertisement featuring Emerson Bakelite radio and phonograph.

Emerson Bakelite radio and phonograph with lid closed.

No name model unknown. Amplifier and speaker box in green and white designed to hold rp-190 record changer attachment. Volume and tone controls mounted in lid. Notice cutouts to accommodate spindle and tonearm. $125-$200. *Courtesy Robert Becker Collection.*

No name model unknown. Amplifier and speaker box in brown and tan designed to hold rp-190 record changer attachment. Volume and tone controls mounted in lid. Notice cutouts to accommodate spindle and tonearm. $125-$200.

Amplifier and speaker box in brown and tan designed to hold rp-190 record changer attachment shown with lid closed.

Harleigh model unknown. Portable brown and tan with tone control. Equipped with rp-190 changer. Single stage amplification, brown knobs on lower right front two-tone grill. $100-$175.

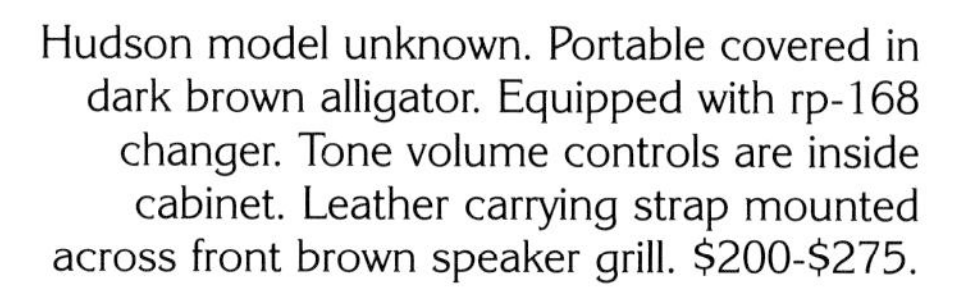

Hudson model unknown. Portable covered in dark brown alligator. Equipped with rp-168 changer. Tone volume controls are inside cabinet. Leather carrying strap mounted across front brown speaker grill. $200-$275.

Majorette model unknown. On left, compac portable in orange and tan. On righ portable in green and white. Each equippe with rp-190 changer. Volume and ton controls on lower right of grill. Single stag amp. $100-$175

Harleigh model unknown. Portable brown and tan with tone control shown with lid closed.

Lafayette model unknown. Two-tone portable white and black with interesting line treatment on lid. Equipped with rp-190 changer. Very compact with black volume control knob on lower right front grill. Single stage amplifier driving a front mounted speaker. $100-$175.

Montgomery Ward model 94GCD864A. Brown metal attachment equipped with rp-168 changer. Company logo is mounted in center of front. $100-$150.

Close-up of Montgomery Ward logo.

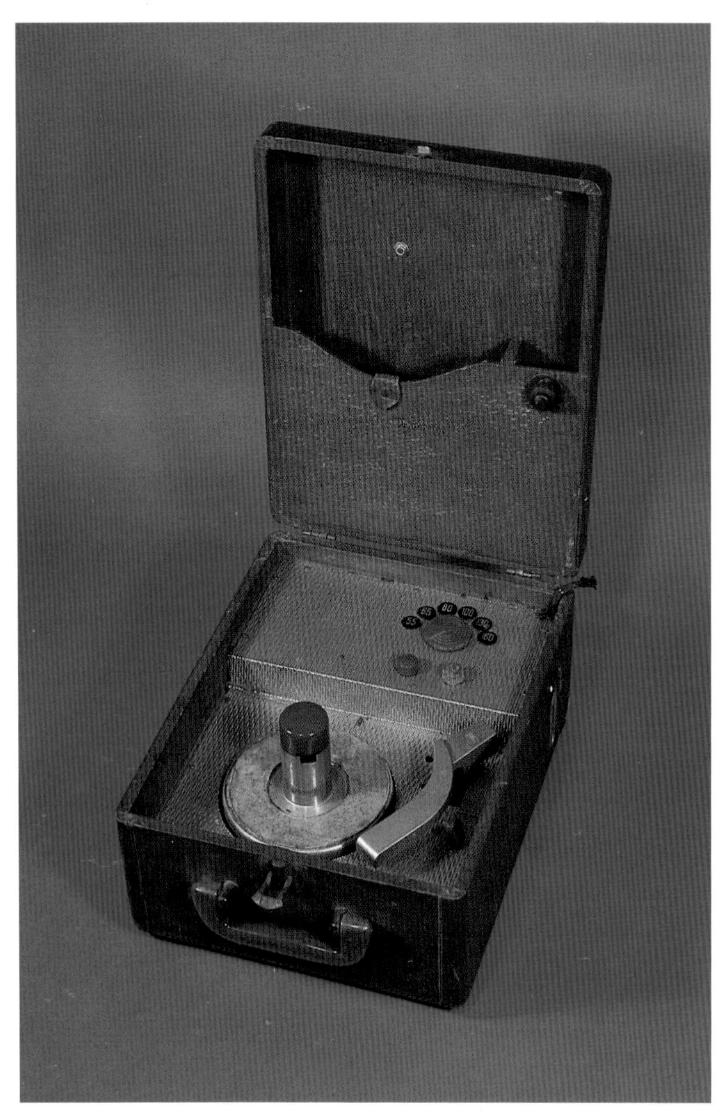

Motorola portable phono with am radio in brown vinyl. Equipped with rp-168 changer. Turntable base and radio covered with anodized gold treatment with radio mounted on step above changer. Record storage is provided in lid. Originally priced at $69.95. $175-$275.

Motorola customized or butchered portable radio phono with padded vinyl covering. Equipped with rp-168 changer. Turntable base and radio covered with anodized gold with radio mounted on step above changer. $175-$225.

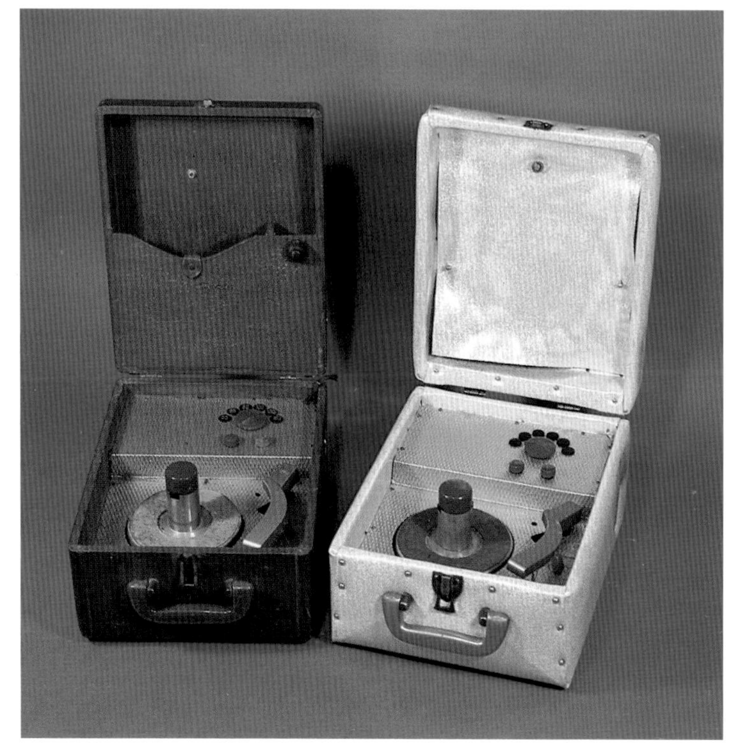

Motorola production and customized phonographs side by side.

Phonola model unknown. Tabletop model in basket-weave tan pattern. Brown tone and volume controls on top right corner. Equipped with rp-190 changer. Single stage amplifier plays through front mounted speaker. Changer base is colorful red. $125-$200.

Pilgrim model unknown. Portable in off-white and gray. Equipped with rp-190 changer. Single stage amplifier plays through front mounted speaker. Volume knob mounted on right side of front mounted grill. $100-$175.

Phono Art model unknown. Portable radio phonograph in charcoal and white. Equipped with rp-190 changer. Five tube super heterodyne plays through front mounted speaker. Nice divided grill treatment. $150-$225.

School Phonograph model unknown. Portable commercial grade cabinet in brown alligator. Detachable lid contains speaker. Equipped with rp-168 changer. Volume control on top front. No product name is present anywhere. Since most schools are equipped to play any size record and speed, this machine is a real oddball. $150-$250.

Silvertone model 6238. Table top model in dark gray with checkered grill in front. Equipped with rp-190 changer. Black volume and tone controls on lower right side. Single stage amplifier drives front mounted speaker. $125-$200.

Sonic model unknown. Two-tone portable white and gray equipped with rp-190 changer. Single stage amplifier drives front speaker. $100-$175. *Courtesy Robert Becker Collection.*

Sonic model unknown. Brown alligator portable case equipped with rp-168 changer. Notice hole in lid to accommodate the spindle. $100-$175. *Courtesy Ray Tyner Collection.*

Sonic model unknown. Two-tone portable white and green equipped with rp-190 changer. Single stage amplifier drives front speaker. $100-$175. *Courtesy Dan Saporito Collection.*

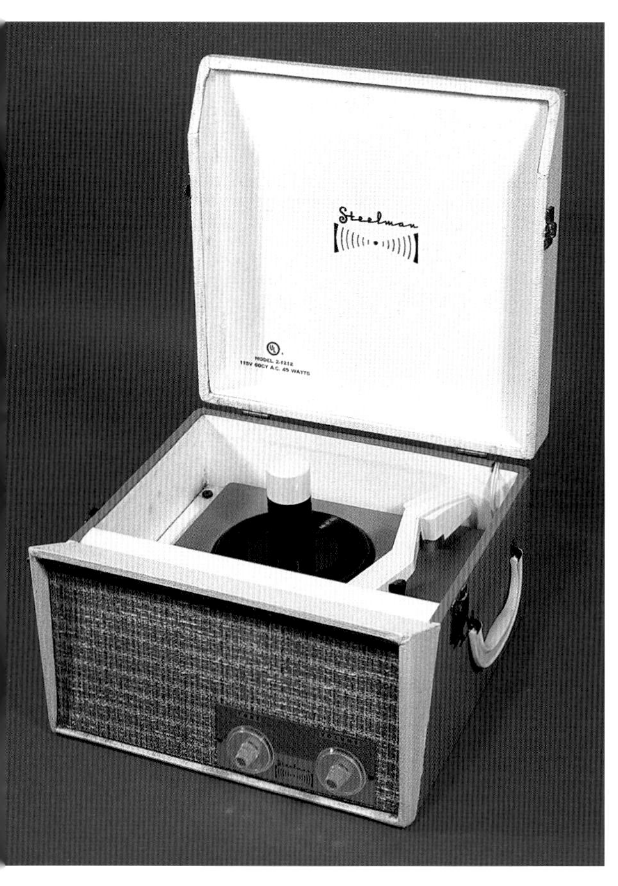

Steelman model 2-1212. Two-tone portable white and peach color with fancy volume and tone knobs mounted on lower right front. Equipped with rp-190 changer. Single stage amplifier drives front speaker covered with colorful grill. $150-$225.

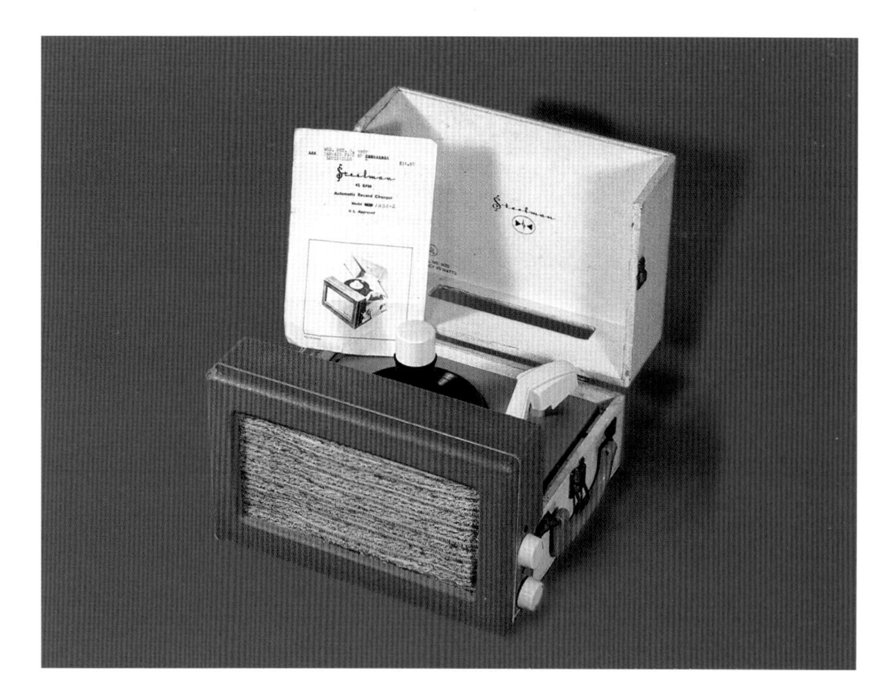

Steelman model unknown. Two-tone portable white and peach color. Tone and volume knobs mounted on lower right side. Equipped with rp-190 changer. Single stage amplifier drives front speaker covered with tweed grill. $100-$175.

Stromberg Carlson model unknown. Metal attachment in brown equipped with rp-168 changer. Slide switches on top for on/off and reject. $150-$250.

Symphonic model unknown. Portable gray and white with tone control. Unique barrel shaped cabinet with spatter pattern on grill. Equipped with rp-190 changer. Single stage amplification clear knobs on lower right of front grill. $200-$275.

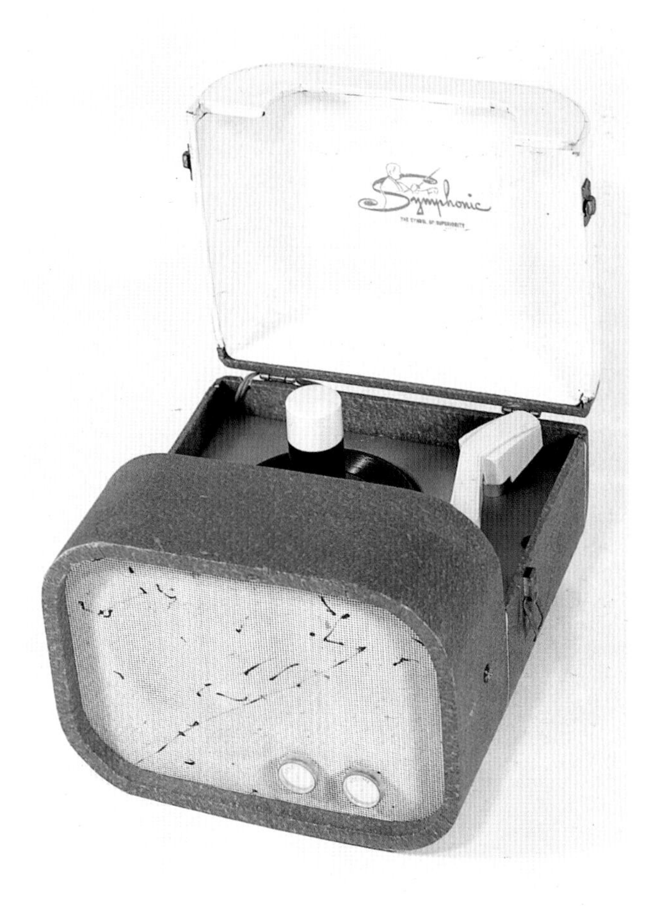

Symphonic model 1045. Wooden table top in mahogany. Equipped with rp-190 changer. Single stage amplification with volume control on lower right front. $150-$225.

Symphonic model unknown. Red and white portable equipped with rp-190 changer. Single stage amplifier drives front mounted speaker in wrap around white grill. Tone and volume controls mounted on lower right front. $150-$225. *Courtesy Robert Becker Collection.*

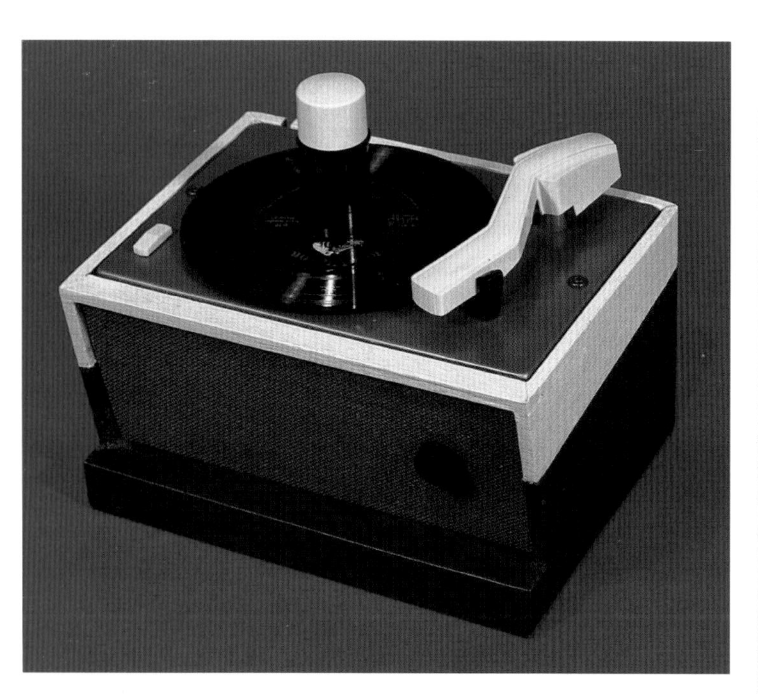

Symphonic model 1245. Table model in gold and white. Equipped with rp-190 changer. Single stage amplification drives front speaker mounted at an angle with step at the bottom of the front. $150-225.

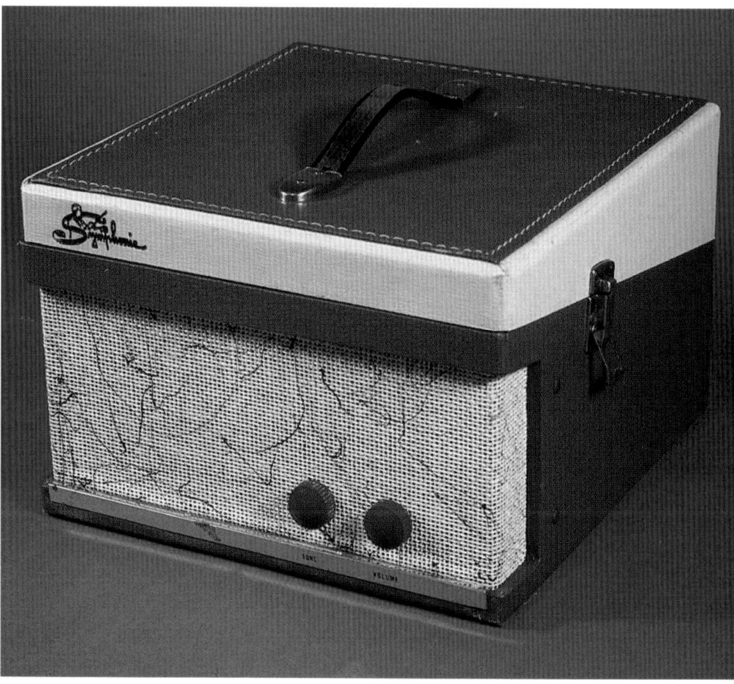

Symphonic red and white portable with lid closed. *Courtesy Robert Becker Collection.*

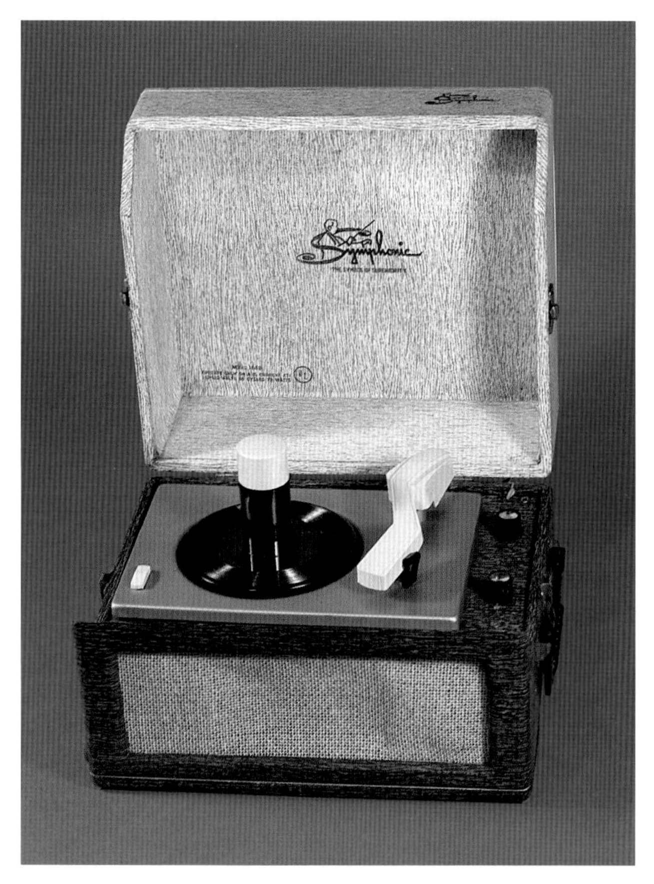

Symphonic model unknown. Gray and white portable equipped with rp-190 changer. Single stage amplifier drives front mounted speaker. Tone and volume controls mounted on right of top side. $100-$175. *Courtesy Robert Becker Collection.*

Teletone model unknown. Two-tone portable in peach and brown. Equipped with rp-190 changer. Single stage amplifier drives front mounted speaker. Volume knob is on front right side of grill cloth. $100-$175.

Symphonic model unknown. Gray and white portable equipped with rp-190 changer. Single stage amplifier drives front mounted speaker. Trapezoid shaped case with volume control mounted on lower right side. $125-$200.

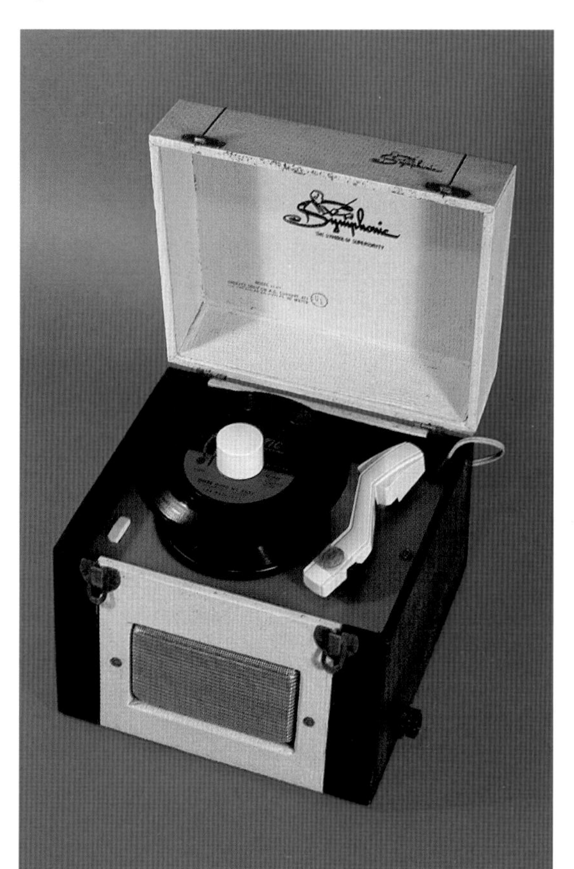

Triton model unknown. Two-tone green portable equipped with rp-190 changer. Single stage amplifier drives front mounted speaker. Tone and volume controls are mounted on right side of front grill. $100-$175. *Courtesy Robert Becker Collection.*

Truetone model D2001A. Brown metal table top attachment equipped with rp-168 changer. Truetone logo appears on top lower left. $100-$150.

Twin Truetone portables, one with the rp-190 changer and the other with the Voice of Music 2-speed changer (see page 105).

Truetone model unknown. Red and charcoal vinyl covered portable equipped with rp-190 changer. Case is trapezoid shaped. Tone and volume controls are mounted on right side of front speaker grill. $175-$250.

Victory model unknown. Brown Bakelite table model identical to RCA Victor 45EY2. Equipped with rp-190 changer and two stage audio amplifier. Tonearm is unique maroon color. Victory logo on top lower right. $225-$325. *Courtesy Dan Saporito Collection.*

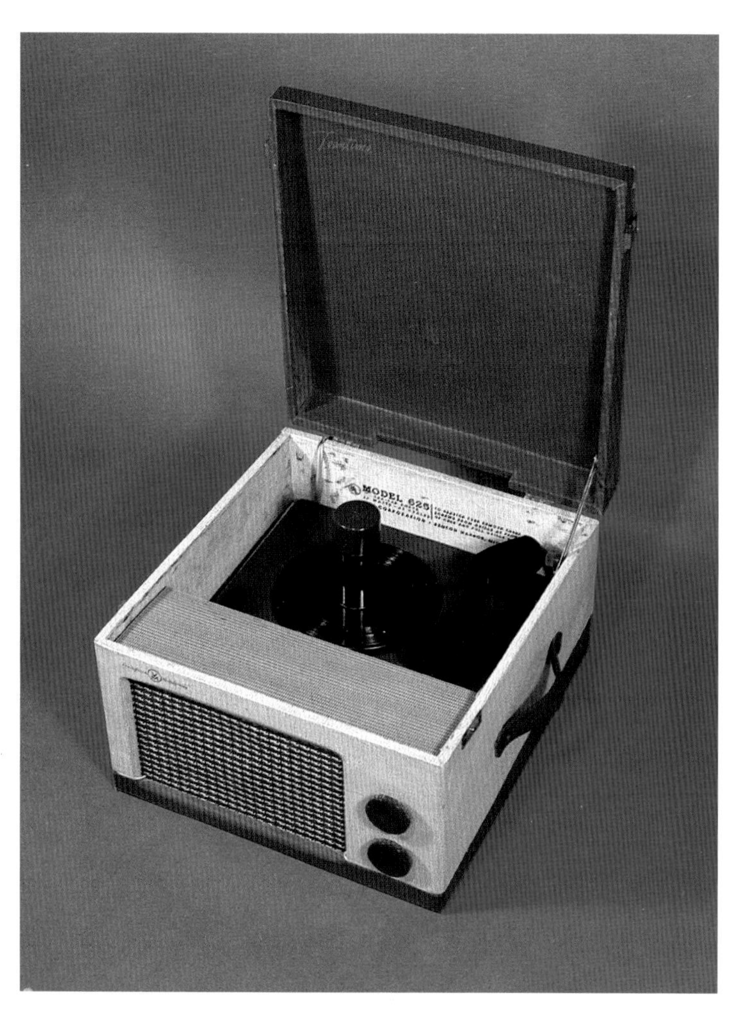

Voice of Music (VM) model 625. Red and gray portable with tone control. Changer base is colorful red. Equipped with rp-190 changer. Single stage amplifier plays through front mounted speaker. Red knobs on lower right of front panel. $125-$200.

Voice of Music (VM) model 625. Orange and white portable with tone control. Changer base is white. Equipped with rp-190 changer. Single stage amplifier plays through front mounted speaker. White knobs on lower right of front panel. $125-$200.

Both Voice of Music (VM) model 625s side by side.

Wards Airline model 948R-2006-A. Wooden table model with radio and phonograph. Equipped with rp-168 changer. Tuning and control knobs are front mounted on lower right. Five tube superheterodyne receiver has single ended output driving front mounted speaker. $200-$275.

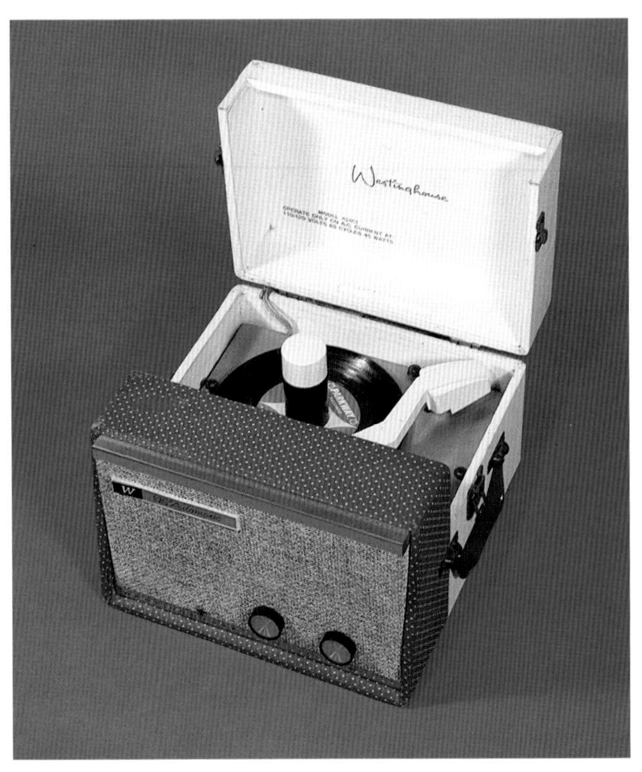

Westinghouse model 42AC1. Portable in red and white has tone control. Equipped with rp-190 changer. Single stage amplifier plays through front mounted speaker. Black knobs with chrome trim on lower right of front grill. $100-$175.

Both Westinghouse model 42AC1s.

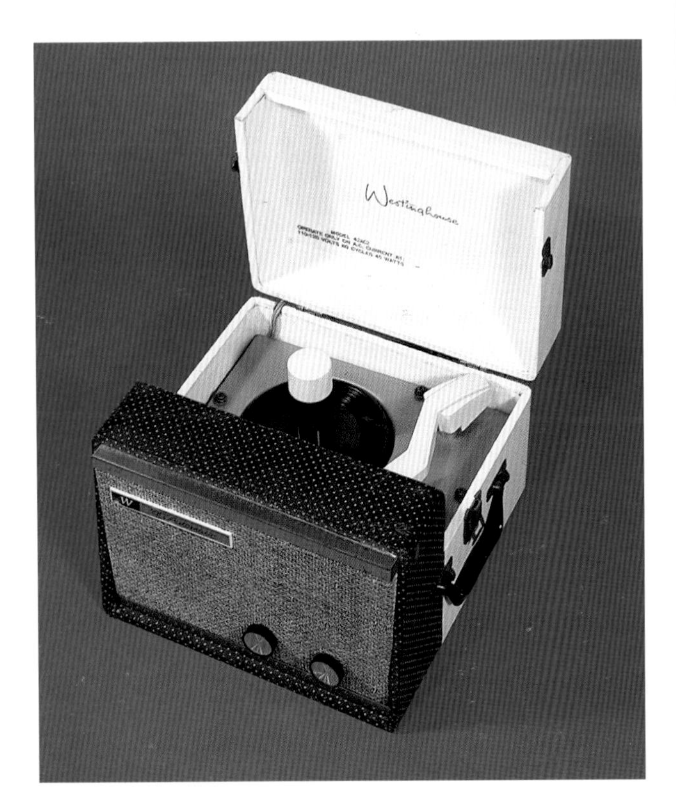

Westinghouse model 42AC1. Portable in gray and white has tone control. Equipped with rp-190 changer. Single stage amplifier plays through front mounted speaker. Black knobs with chrome trim on lower right of front grill. $100-$175.

Zenith attachment model S14020. Bakelite table model equipped with rp-168 changer. Scalloped styling on all sides of the cabinet. Two push buttons mounted on the front for reject and on/off. $125-$175.

Zenith model unknown. Bakelite table model equipped with rp-190 changer. Six tube superheterodyne receiver drives five-inch front mounted speaker mounted behind tuning dial. $250-$350.

Close-up of Zenith attachment model S14020, showing use of two push buttons.

Zenith Bakelite twins, one with rp-190 changer and the other with Twin Seven changer (see page 108).

Chapter Twelve

Oddities

Not to be outdone by RCA Victor, some companies decided to design their own machines to handle the new 45 rpm records. In most cases the machines were able to play records of other speeds that were the same seven-inch size as the 45 rpm records. None of these lasted more than a couple of years in production, but they are interesting nevertheless.

Admiral Radio/Phonograph Two-Speed

This Admiral radio/phonograph two-speed plays seven-inch records. The record changer is designed to play 45 rpm and 33 1/3 rpm seven-inch records. The large center spindle pulls out and a skinny spindle is inserted to handle the 33 1/3 rpm records. At the end of the stack the last record repeats until the unit is turned off.

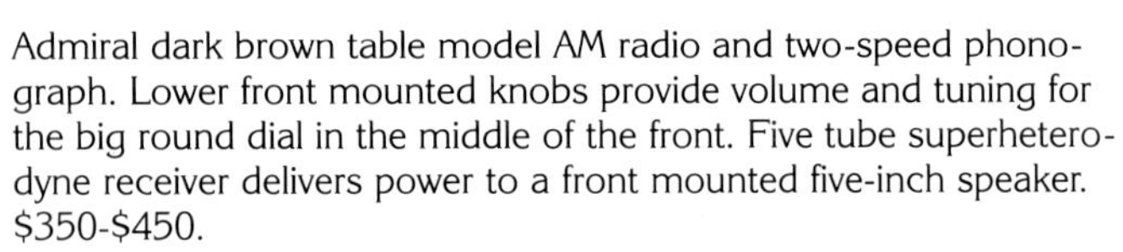

Admiral dark brown table model AM radio and two-speed phonograph. Lower front mounted knobs provide volume and tuning for the big round dial in the middle of the front. Five tube superheterodyne receiver delivers power to a front mounted five-inch speaker. $350-$450.

Admiral radio and two-speed phonograph plays 33 1/3 rpm seven-inch records by removing the larger spindle.

Voice of Music Two-Speed Record Changer

This unusual changer plays 45 rpm and 16 2/3 rpm seven-inch records. The fidelity on these slow moving records (16 2/3 rpm) was not the best, so most records featured talking books. They were not very popular and fell out of favor quickly. The record changer is one of the only seven-inch record changers that shuts itself off after the last record is played. The machine accomplishes the shutoff by probing for the next record on the spindle before it drops it and plays it. If the tonearm does not contact a record during the probing, the player automatically shuts itself off. VM put this changer in their own units and also licensed it out to other manufacturers—Arvin and Teletone were two such licensees. VM and Arvin also designed stereo 45 rpm phonographs.

Truetone portable model in red and black with angled sides. Single stage amplification, red volume and tone knobs on lower right of silver front grill. Voice of Music record changer plays seven-inch 45s and seven-inch 16 2/3 rpm records. Automatic shutoff. $200-$300.

Seven-inch 16 2/3 rpm record featuring the Bible.

Dean portable model in blue vinyl. Single stage amplification. Gray volume and tone knobs on lower right of tweed front grill. Voice of Music record changer plays seven-inch 45s and seven-inch 16 2/3 rpm records. Automatic shutoff. $175-$250.

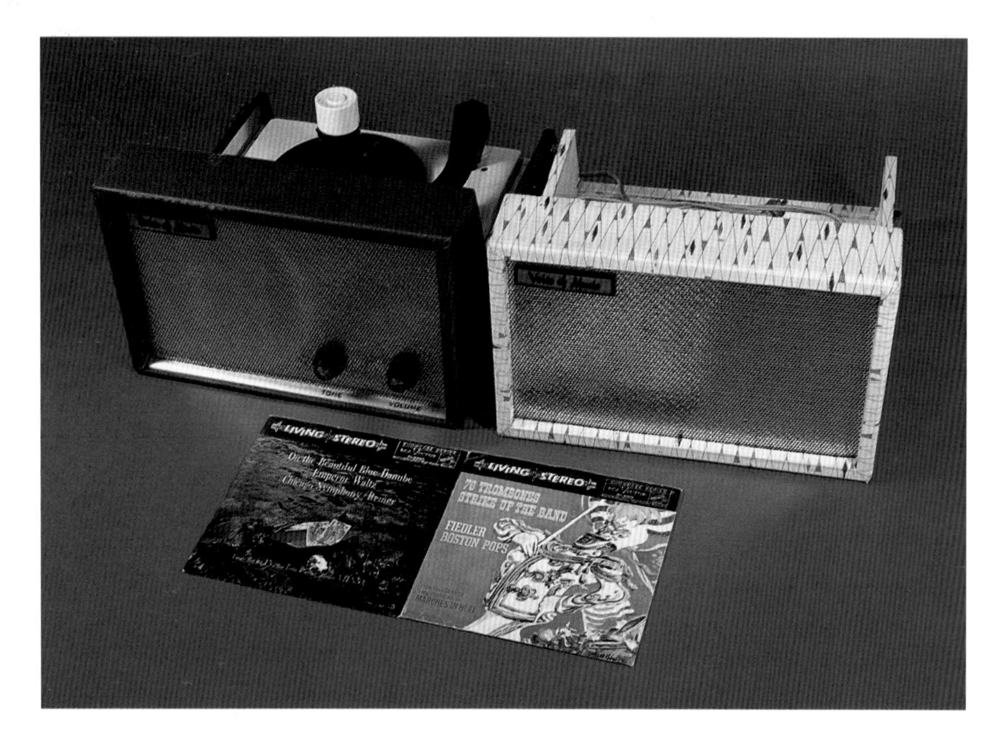

Voice of Music (VM) model 301 portable stereo in silver and black. Dual single stage amplifiers driving front mounted five-inch speaker and front mounted five-inch speaker in lid. Black volume knobs on lower right of silver front grill. Record changer plays seven-inch 45s and seven-inch 16 2/3 rpm records. Automatic shutoff. Late '50s stereo 45s are displayed as well. $300-$400.

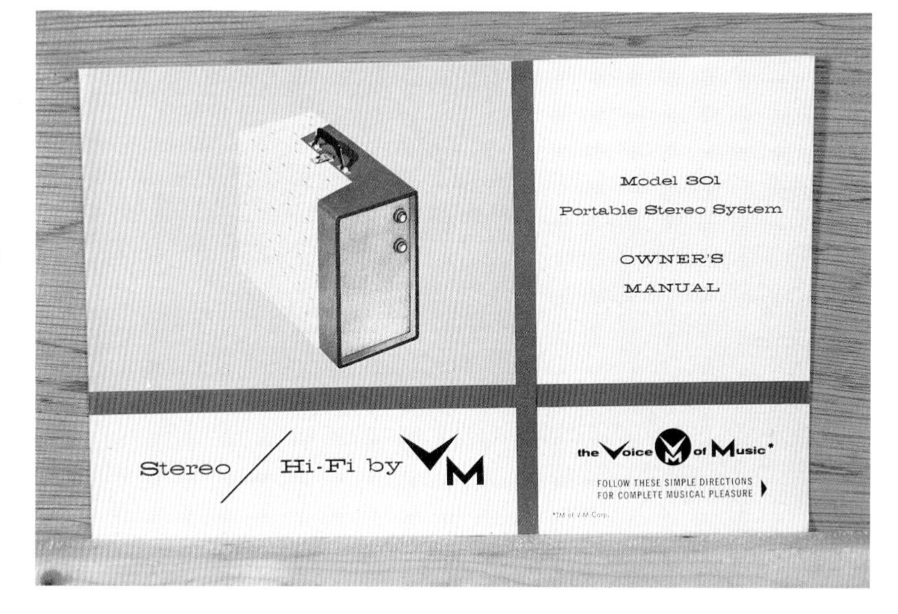

Owner's manual for Voice of Music (VM) model 301 portable stereo.

Voice of Music (VM) portable stereo model 301 in silver and black shown packed up and ready to be carried to the next party.

Arvin portable stereo model in two-tone brown and tan. Unit folds out like opening a clam with a speaker mounted in each side, changer mounted in left clam shell, and record storage in right clam shell. Dual single stage amplifiers drive the speakers. Volume, tone and balance controls are mounted on the lower right front of the left clamshell. VM record changer plays seven-inch 45s and seven-inch 16 2/3 rpm records. Automatic shutoff. $300-$400.

Arvin portable stereo model in two-tone brown and tan shown packed up and ready to go.

Webster Chicago Two-Speed Models 760 and 762.

This record changer plays 45 rpm and 33 1/3 rpm seven-inch records. It does not use a fat spindle for the large-hole 45 rpm records. Instead, the user must insert a 45 rpm adaptor into each record so the small spindle will support the records. Webster-Chicago was also in the business of manufacturing the adaptors, so the design of this changer was very good for business. No balancing apparatus is used, and the records are held on the spindle by a tripod of metal tongs. In order to drop one record at a time during the reject cycle, a small rubber washer mounted on the spindle is compressed during the cycle, and it prevents the remaining records from falling. Unfortunately, with age this rubber washer usually rots and falls apart. A replacement part is no longer available, so the restorer must find something to take its place. Getting the records to drop correctly is not easy.

Webster Chicago model 760-1. Dark brown table model in Bakelite with lid. Knobs are located on lower right front. Record changer handles seven-inch 45s and seven-inch 33s. $300-$400. *Courtesy Joe Centanni Collection.*

Webster Chicago model 762. Dark brown portable in vinyl with lid, shown playing seven-inch 33 rpm records.

Webster Chicago model 762. Dark brown portable in vinyl with lid, shown playing seven-inch 45 rpm records.

Webster Chicago model 762. Front view with lid closed. Volume and tone controls are mounted on upper front of right side. Model number prominently displayed on top of front grill. $250-$350.

Zenith Twin Seven

The title itself is revealing—this record changer has two turntables! One table handles 45 rpm records, the other plays seven-inch 33 1/3 rpm records. One miniature cobra pickup arm is located between the two turntables, so that it can serve the selected table. Records on one turntable partially overlap the other turntable, so only one table at a time can be loaded with a stack of records, i.e., only one type of record can be played at a time. There are two different set positions for the pickup arm and it must be set before playing a stack of records. Both turntables turn at all times while the records are playing. It is truly a strange sight. This changer can also be found in Bakelite table models and consoles.

Zenith Twin Seven. Designed to fit in console slide out drawer. $300-$500. *Courtesy Bob Havalack Collection.*

Zenith Twin Seven. Dark brown table model AM radio and two-speed phonograph in Bakelite. Knobs are mounted in lower front on each side of a large radio dial with speaker mounted within it. Five tube superheterodyne receiver provides decent sound through five-inch speaker. Shown set up and playing 45 rpm records. $400-$600.

Zenith Twin Seven playing 33 rpm seven-inch records.

Silvertone Manumatic

As the name implies, some of the records are played automatically (16 2/3 rpm and 45 rpm) and the others are played manually (33 1/3 rpm and 78 rpm). The record changing mechanism resembles that on the RCA Victor rp-168 in that the entire reject cycle is accomplished in one revolution of the turntable. This occurs because the reject cycle uses a cam mounted under the turntable platter. There is no automatic shutoff. When the 45 spindle is installed, it rotates along with the turntable.

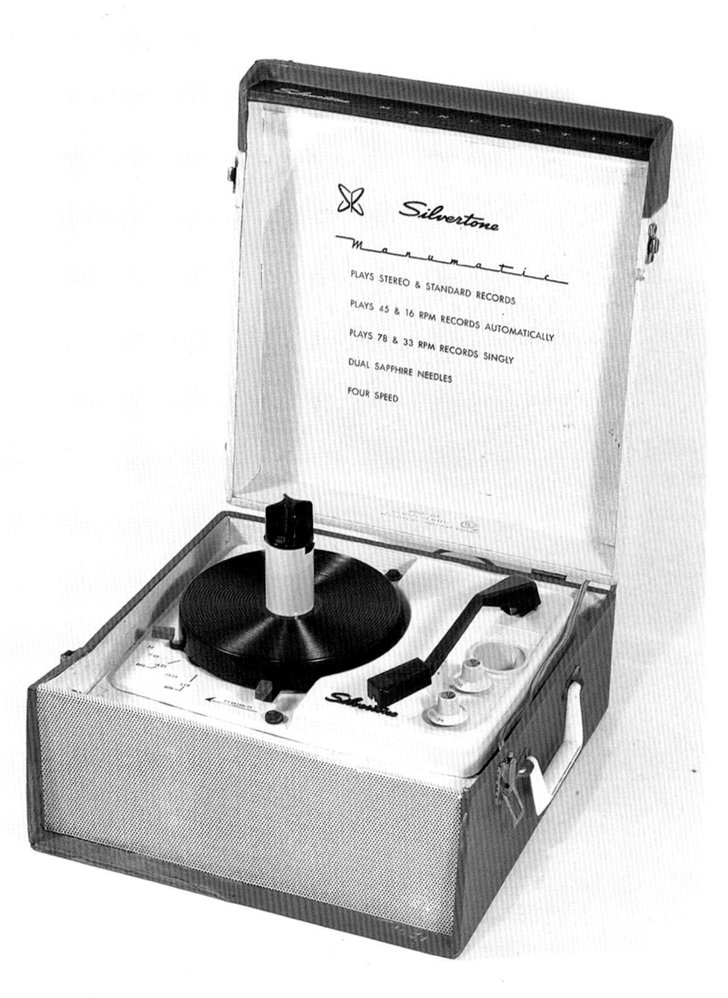

Silvertone Manumatic record player. Available in various vinyl covered portable designs. Audio amplifier drives four-inch front mounted speaker. $150-$225.

RCA Victor Counters with Their Two-Speed

As if the above units were not strange enough, RCA Victor introduced a two-speed unit that played a stack of 45 rpm records or one 33 1/3 rpm LP. It was a bare-bones design with only one stacking spindle, for the 45 rpm records. When that spindle was removed, the stub underneath it was used for the 33 1/3 rpm LP. There was no automatic shutoff. This design was introduced in the later 1950s when multispeed changers were available everywhere. As a consequence, not many of these two-speed 45 rpm units were sold.

The changer was available in attachment form and in two amplified versions, with the more powerful unit called "Deluxe." It was available also in colorful plastic or vinyl-covered portables.

RCA Victor 9E attachment. Table model in black and white plastic. Can be plugged into any radio with a phono plug. All RCA Victor radios from this era were equipped with this feature. $125-$175.

RCA Victor two speed. Available in gay pastel-colored plastic (9-ED-1). Volume control is on lower right side. One tube audio amplifier drives front mounted four-inch speaker. $225-$325.

RCA Victor Deluxe two speed. Shown in gay pastel-colored plastic (9-ED-2). Volume and tone control are on lower right side. Three tube audio amplifier drives front mounted four by six inch speaker. $225-$325.

RCA Victor portable two speed. Vinyl covered model 9-ED-31. Record changer is designed to play a stack of 45s automatically or to play one LP. There is no automatic shutoff. The large center spindle pulls out and leaves a stubby spindle for the LP. $150-$225.

Custom Cabinets

Some enterprising individuals built their own custom cabinets for their phonographs. As the photographs show, some cabinets were slapped together with little regard to how they looked, while others resulted from painstaking efforts to make an attractive cabinet that also provided record storage compartments.

My favorite is the "modified suitcase." It was obviously designed to provide music at a party. Equipped with dual turntables, lights, and controls, it is truly the precursor to the first party "DJ". The creator of this machine exercised particular skill in mounting the turntables so they would fit under the lid. Nice job!

Cabinet for phonograph attachment made of old produce crates. Partial color ad can be seen inside.

Cabinet made to house 9EY3 or 45EY. Also includes home-made amplifier. Definitely an amateur woodworker but he did put a finish on the cabinet.

Cabinet with beveled corners shown with lid closed. A fancy hinge and latch work together to cover or uncover the speaker area in the front.

Modified luggage carrier houses two 45J2s on a wooden mounting. Also contains a light for each changer and controls for cueing and power. Cable carrying compartment has been added to the top of the piece. Now all you need are some disco 45s. $250-$350.

Cabinet houses 9EY3 or 45EY and has ample storage for records. Judging by the beveled corners and hardware, this was done by a pro or semi pro woodworker.

Custom treatments can also be applied to the phonograph itself; an example is the patriotic color scheme on this 45J2 attachment. Some white spray paint and modern window decals spruce up a 45EY2, as well.

Patriotic model 45J2 dressed up with painted white cabinet, blue motorboard, and red arm and turntable. $75-$120.

White painted 45EY2 can be decorated with window decals that are now sold in dollar stores. As the seasons change you can customize the player with different decals—here it is decked out for the Christmas season. Decals go on dry or with water and come off easily. $200-$300.

Same painted 45EY2 with Halloween decals featuring Mickey Mouse and friends.

Single Play

Here is a new-in-the-box single-play 45 record changer. As equipped, it would go into a console unit. But who would want to play 45s singly in a console?

Replacement single-play record changer, new in the box. Designed to be used in any of the consoles RCA Victor made in 1949 or 1950. $150-$200.

After Market Amp

Here is an rp-168 attachment that has had the bottom removed and an amplifier inserted between the Bakelite case and the bottom. Speaker fires sound out of the bottom.

Possibly an aftermarket amplifier designed to fit under the Bakelite case of an rp-168 attachment. $150-$200.

Table Top Jukeboxes

Several companies—Chicago Coin's Hit Parade, Music Mite, and Ristaucrat are representative—manufactured compact table top jukeboxes for small businesses or home use.

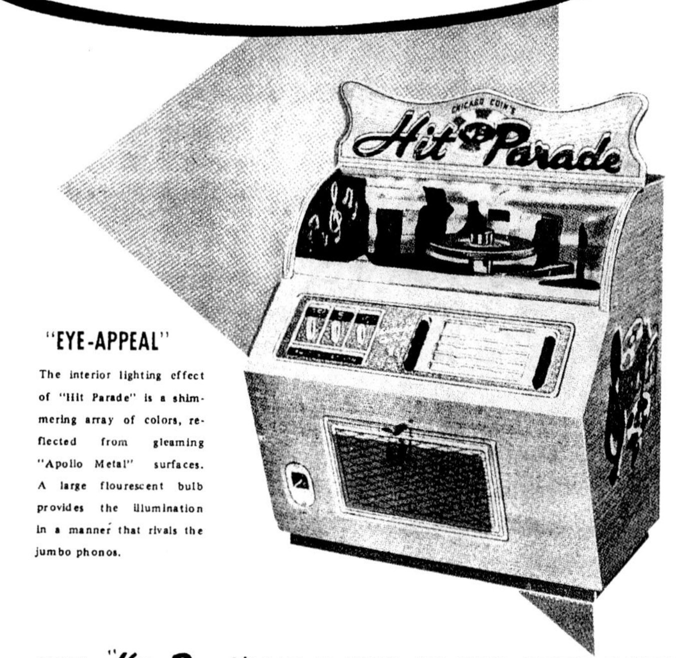

Hit Parade advertisement for table top jukebox, 1951.

Williams "Music Mite" advertisement for table top jukebox, 1951. This is the only table top juke that uses the RCA Victor record changer.

Looking at the Ristaucrat line, it is clear that one could buy a changer attachment that would plug into any amplifier or an AM radio/45 rpm phonograph. They were available in both light and dark woods. The novel feature of these machines is their ability to pick up the records automatically after they have been played. In other words, the changer repositions the records on the spindle so that the stack can be re-played.

Ristaucrat table model record changer attachment. Must be plugged into amplifier. Has novel restacking feature. Available in dark and light woods. $175-$250. *Courtesy Dan Saporito Collection.*

Top view of Ristaucrat table model record changer attachment. *Courtesy Mark Shoenthal Collection.*

Ristaucrat table model AM radio and phonograph. Will play up to eighteen songs and then automatically switch to radio. Has novel restacking feature. Available in dark and light woods. $250-$400. *Courtesy Robert Becker Collection.*

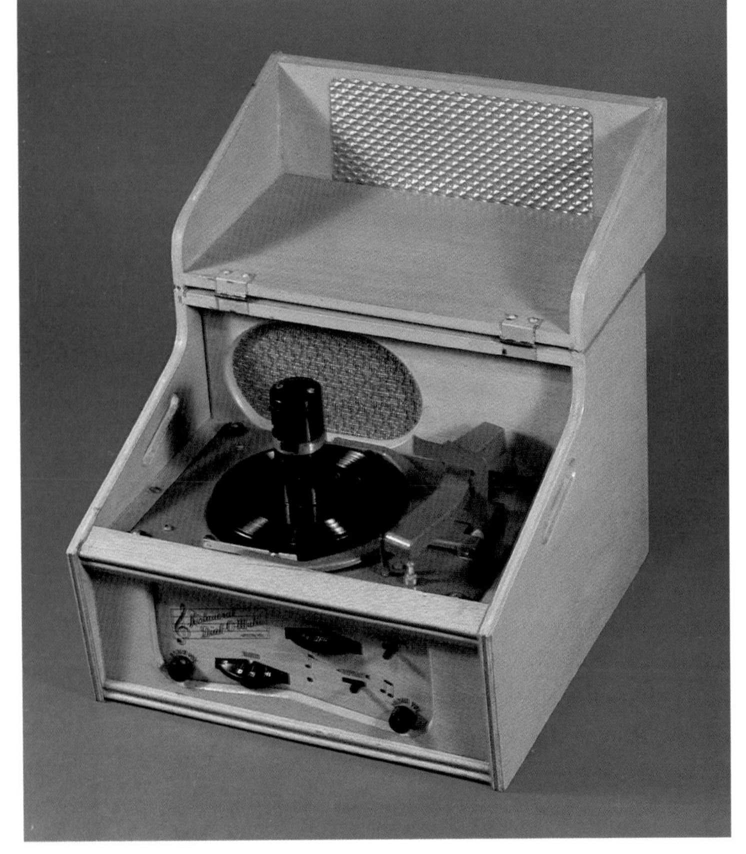

Ristaucrat table model AM radio and phonograph shown with lid closed. Speaker is mounted in rear of changer area. Sound waves travel through changer compartment and come through vents in front of lid. *Courtesy Robert Becker Collection.*

Moving into the small business arena, one could purchase the non-selectable model that for five cents would play the next record on the stack. The following year the selectable model arrived, giving the user the ability to choose any of twelve selections. The reader might naturally ask, "What if record 6 is playing and now I want to choose record 5?" Not a bit of a problem. After record 6 completes its play, the tonearm will move away from the records, and the spindle will drop the remaining records one at a time until the complete stack is on the turntable. Then a horseshoe will pick up the stack from the bottom and move it to the top of the spindle. The spindle will again start dropping records until it drops record 5, at which point the tonearm will descend onto the record. Tedious, perhaps, but it does work and is fun to watch.

Selectable table model jukebox. Choose from twelve selections. Has novel restacking feature. Available in oak finish. This unit was restored in red, which is not an original color. $600-$1000.

Non-selectable table model jukebox. Putting a nickel in the slot plays the next available record, whatever that may be. No labels, has novel restacking feature. Available in oak finish. $400-$700.

Selectable table model jukebox lifting stack of records.

Chapter Thirteen

Optional Accessories

RCA Victor offered some optional accessories, which included carrying cases, extension speakers, and an AC power supply for their battery-powered portable unit. Such accessories as record changer covers, record carrying cases, and "stackers" were also available on the after-market.

Optional wooden case for early RCA Victor attachments (rp-168).

Optional soft vinyl case for RCA Victor models 9EY3, 45EY, and 45EY1.

Soft cases were available for many Bakelite models. At left is a case for model 9EY3. On the right is a case for the 7EY series.

Prototype advertisement sent to RCA managers showing 9EY3 with optional carrying case. *Courtesy Camden County Historical Society.*

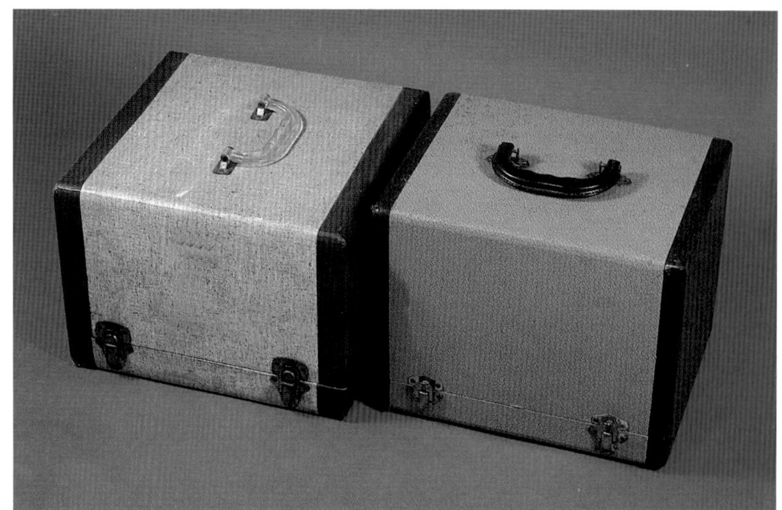

Optional vinyl covered wooden cases were also available. At left is a case for model 45EY2. At right is a case for 7EY series.

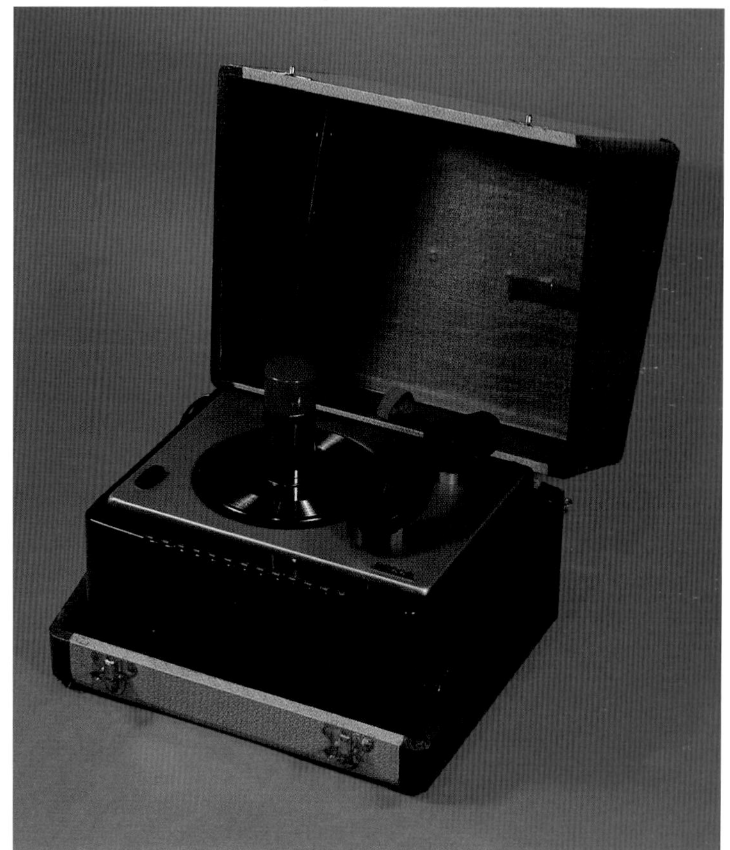

Optional vinyl covered wooden case holding 45EY2 clone sold by Decca.

Optional soft vinyl case holding 7EY1.

Optional vinyl covered wooden case holding 7EY1.

Aftermarket record case featuring Pat Boone. *Courtesy Robert Becker Collection.*

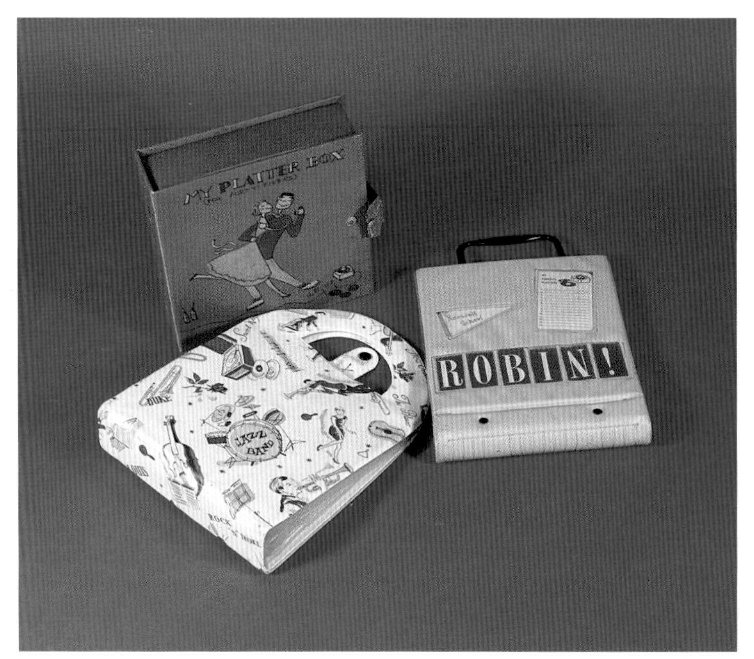

An assortment of aftermarket record cases. *Courtesy Robert Becker Collection.*

An assortment of "Tune Tote" record carrying cases. *Courtesy Robert Becker Collection.*

Optional external speaker in oak wood finish.

Model 6BY4A shown with optional AC power supply. Connector looks like a 7 pin tube socket.

Aftermarket plastic cover for 45EY2 phonograph.

Optional dealer display featuring a rotating carousel mounted on the 45 spindle. A special record called the "Whirlaway Demonstration Record" plays and describes the new 45 system. *Courtesy Camden County Historical Society.*

Record player selector switch designed to be installed in your existing phonograph or radio. It allows your new 45 rpm attachment to play through the system.

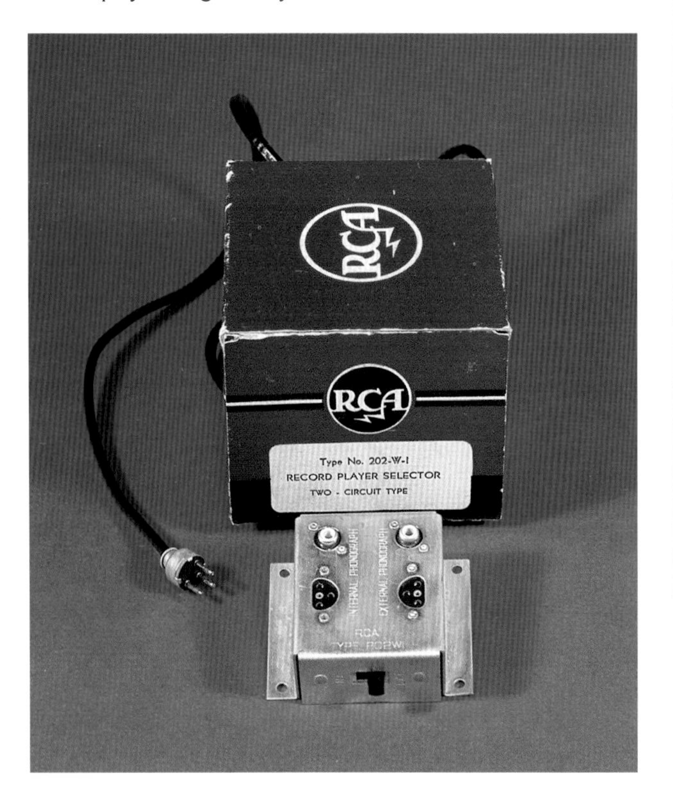

Plastic record "stackers" were sold on the aftermarket. Notice that the colors match many of the early RCA Victor colored vinyl records. *Courtesy Bob Havalack Collection. Photo by Tim Fabrizio.*

Plastic record "stacker" holding some 45s. *Courtesy Bob Havalack Collection. Photo by Tim Fabrizio.*

Plastic record "stackers" stacked together. *Courtesy Bob Havalack Collection. Photo by Tim Fabrizio.*

Chapter Fourteen

Repair and Restoration

This section will describe how to completely rebuild and restore the phonograph. Unlike repairing a problem in the shortest amount of time like a repairman would, this process will take hours to accomplish but the outcome will be a unit that looks and plays like new. It is recommended that this process not be done in one sitting, but rather in two or three sittings.

General Restoration Tips

When you start working on one of these units you must first ask yourself, am I repairing or restoring? If you are restoring, the first order of business is to take the item completely apart so that the cabinet is empty. Now the cabinet can be cleaned thoroughly without other parts in the way. Areas that should not be overlooked are where the record changer sits, any corners that have built up grime, and areas where controls are located. One complication with this is paper glued to the inside or bottom of the cabinet. Trying to remove such old paper can be tricky and may lead to rips and tears. If the paper can't be removed, you'll want to avoid getting this part wet. Other options are to put clear tape over the paper or spray it with clear-coat. Avoid over-spray by masking the area to be sprayed, in order to avoid getting spray all over the interior of the cabinet.

Cleaning of Record Changers

If you are doing a complete restoration, I recommend the following steps to clean and lubricate the record changers. Moderate strength cleaners like Formula 409 or Whestley's Bleche Wite can be used to clean the metal motorboard. The motorboard is the painted metal housing in which the pickup arm is mounted and upon which the turntable sits. All parts should be removed before cleaning (refer to Disassembly). A stronger cleaner like Castrol Superclean will work fine on older models (1949 through 1954) but will damage clear-coat finishes that are used on some of the later models like model 7EY1 or 7EY2 (refer to Chapter Seven for examples of the 7EY series). Once the motorboard is clean, apply Novus #2 if you wish to polish to a high luster. Novus #2 can also remove some stains that have not penetrated too deeply into the paint. I have found that Novus #2 can be used on many surfaces successfully besides plastic. Clean moving and pivoting parts with denatured alcohol. The more stubborn pieces can be cleaned with lacquer thinner but be careful about providing enough ventilation. Clean plastic pickup arms and turntables with soap and water. Then buff them up with Novus #2. Do not use chemical cleaners, especially on the plastic pickup arm. Depending on how fussy you want to be about the restoration, the metal parts on the underside of the record changer can look like new if you follow the procedure in the video *RCA 45EY2 Phonograph Restoration with Paul Childress*.[254] Scrub each metal piece with Gojo cream hand cleaner and rinse with water. After drying, apply polishing compound (Dupont #7) and buff out to remove any oxidation. This scrubbing procedure may need to be applied more than once. Once you are done the piece will look like new.

Shown are many of the products used by the author to restore phonographs.

Motors

RCA Victor contracted with different manufacturers to ensure there would be no shortage of motors during assembly. The companies that provided most of the motors were Alliance and GI. Several different voltages were used to accommodate the different circuit configurations like 85 volts, 90 volts, and 115 volts. The third motor pictured was only used as a last resort and was referred to as the "toy motor." Ironically the "toy motor" seems to work whenever I find them while the others tend to run slow.

Alliance, main supplier of motors for 45 rpm changers.

GI, alternate supplier of motors for 45 rpm changers.

Last resort emergency supplier of motors for 45 rpm changers. Referred to as the "toy motor" (supplier unknown).

Rubber Parts

Rubber is used to isolate the motor from the motorboard and to isolate the motorboard from the cabinet. These parts disintegrate over time and now that they are approaching fifty or more years old, it is a very good bet that the rubber is damaged. The motor grommets are still available at some electronic supply stores because they were used on most of the record changers into the 1980s. The cone shaped rubber used to isolate the changer from the cabinet is another story. These are impossible to find. If the cone is only partially squashed, adding one or two washers to the bottom of the cone can provide the needed separation between the changer and the cabinet. The cushioning factor may not be exactly the same but it does work.

Suspension parts consisting of three cones to suspend the motorboard and three bushings to support the motor.

The rubber part that actually provides the connection between motor and turntable is the idler wheel. This part must have the proper dimensions to rotate at 45 rpm. It must be supple enough and have enough grip to turn the turntable evenly and without slippage. When the rubber gets old it becomes hard and does not have enough gripping power. This causes the mechanism to stall during the reject cycle. This is when the machine is under heavy load. The other common problem with the idler is flat spots. The design of the record changer keeps the rubber idler in constant contact with the motor shaft and the turntable. If the phonograph is put up in the attic or basement and left there for years, the rubber will flatten because it is contacting the other parts. Once there are flat spots and the rubber hardens with age, it is next to impossible to remedy the situation. Back in the early 1950s you could go to your local RCA Victor parts dealer and buy a new idler like the one shown for eighty cents. Later on I will describe the methods I use to bring some of the old idlers back to life.

New old stock idler wheel.

The rp-190 45 changer uses a rubber cycling cam. While this provides nice smooth record changing motion, it becomes a problem when the rubber deteriorates. The reject cycle will start slipping and ultimately the machine will not cycle at all. Early cycling cams had a generous amount of rubber along the edges. Later ones were cost reduced and had a much thinner layer of rubber.

Cartridge Replacement

Originally most of the phonographs were equipped with a crystal cartridge with a low, medium, or high output depending on the amplifier it was equipped with. In the mid 1950s a ceramic cartridge was offered on the better fidelity units. The only replacement that is available today is an Astatic replacement for the ceramic cartridge. It is model 51-1. These are occasionally found because they are still used in certain 45 rpm jukeboxes. The 51-1 sounds pretty good but does not track stereo records well at all. The original crystal cartridge can be rebuilt but I have had mixed results with this. Sometimes the rebuilds sound tinny. Some collectors are using Power Point cartridges that were used on Califone phonographs in public schools throughout the country. The slide-in replacements are still available from Pfansteil distributors. The problem is finding the body that holds the slide-in replacements. Many an old Califone has had its tonearm extricated for this reason. Another alternative is to use the original casing to install a more modern cartridge like the Sharp model 146. These are available now on the Internet and work very well. The last way to deal with this is by installing a more modern cartridge with a glue gun. If this job is done carefully, the results can be quite good. The purists in the group will have nothing to do with this however.

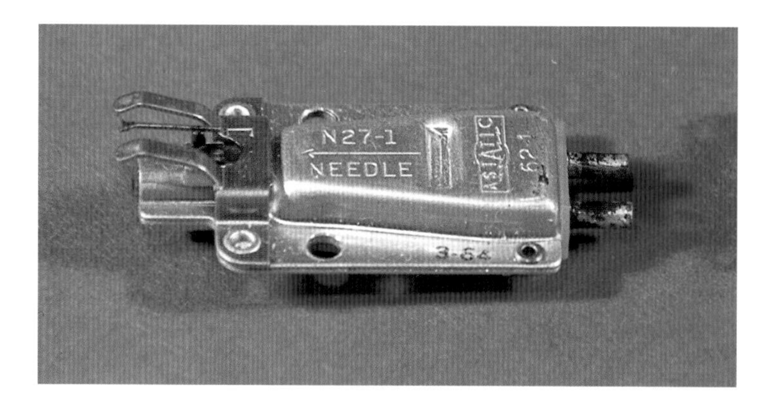

Crystal cartridge used on most 45 rpm phonographs.

Cycling cams. The early one on the left has much more rubber than the later one on the right.

Sonotone ceramic cartridge used on higher fidelity models.

Sharp 146 replacement cartridge integrated into existing crystal cartridge. *Courtesy Paul Childress Collection.*

Sharp 146 replacement cartridge integrated into existing crystal cartridge using original connecting pins. *Courtesy William Bosco Collection.*

Power Point cartridge found in many school record players.

Rp-168 Record Changer

The rp-168 changer was the first to be introduced by RCA. The machine had been designed in the early 1940s. Even though it was designed to be simple (fewer moving parts, no automatic shutoff), it was very well made. A heavy brass turntable was used in order to minimize wow and flutter, a mute switch was intended to silence any reject cycle noise, and models that were free-standing phonographs complete with amplifiers and speakers would be connectorized so that repair was simplified.

Improvements

The biggest problem with the rp-168 changer was the cycling mechanism. Because it was so fast, the tonearm would usually bounce as it made contact with the next record in sequence. In order to alleviate this problem, a damping piston was added so that it engaged when the tonearm was falling to the record.

The early batches of spindle caps were found to be prone to cracking, so the cap was redesigned to be thicker. This also meant a change in the mating surface of the spindle where the cap was mounted.

The record-dropping mechanism inside the center spindle worked well until an out-of-specification record appeared in the stack. Since gearing operated the system of shelves and knives that functioned both to retain the stack of records on the spindle and to drop each single record in sequence, the out-of-spec record, with its aberrant center hole of abnormal thickness at the center hole, would cause the shelves and knives to bind and freeze up the record changer. The knives would get stuck in the record's center hole. The bottom record in the stack could not then be removed easily, and frustrated customers would try to pry it off, breaking the record and/or the mechanism of the spindle assembly. A new record-dropping mechanism was introduced in 1950 that would yield under the binding condition, allowing the spindle to be manually turned past the reject cycle, thereby freeing up the mechanism.

The last problem with the rp-168 changer began to be evident as the fidelity and loudness of 45 rpm records improved and increased respectively. Some of the new records (those on the Liberty and Dot labels were especially notable here) with their deep and loud bass passages, would cause the stylus to jump out of the groove and skip. This phenomenon may also have been aggravated by the cartridge and stylus becoming older and less compliant. In any case, many consumers resorted to placing pennies and nickels on top of the tonearm in order to add weight to the stylus and keep it in the grooves. A number of buffalo head nickels have been found on players unearthed by the author!

Lest the reader begin to get an erroneous impression of the utility of the rp-168 changer, it is necessary to point out that this first in the series of RCA's development of the 45 rpm record changer was very durable and typically enjoyed a long life. I had one in my living-room from 1950 through 1957, when it was replaced by another model. The rp-168 changer was produced from 1949 through sometime in 1951.

Cost-Cutting Measures

Removing the tonearm rest from the top of the Bakelite simplified molding of the Bakelite case. Early phono attachments—the playback machine that plugged into a radio and used the radio's amplifier and speaker—were provided with an on/off switch that incorporated a volume control. In later designs, the switch operated only as on/off, and volume was changed at the radio's volume control.

Disassembly

When taking something like this apart it is important that you keep track of all the parts. Use a small bowl or container to hold the smaller parts like nuts, bolts, springs, etc.

Remove the bottom panel from the cabinet.

Unsolder three thin colored wires from the muting switch. Make a diagram so you know where the wires will go when you have to solder them back later. These are the wires that go through the tonearm and connect to the cartridge.

Loosen the nut and bolt that hold the clamp and trip lever (58) to the bottom of the tonearm shaft (pivot arm) and remove the clamp and trip lever. If it still won't come off use a screwdriver blade to spread the clamp apart. Once the clamp is removed, slide the tonearm out from the top of the record changer.

If the unit is equipped with an amplifier, disconnect the 3-prong plug at the amplifier if so equipped. If there is a black 2-prong connector on the motor wires, disconnect it also.

If the record changer has a metal motorboard, remove the three suspension bolts and springs from the bottom and pull the changer out from the top of the cabinet.

Remove the mounting screws holding the record changer to the motorboard.

If the record changer is mounted in a Bakelite cabinet (doesn't have a motorboard), remove the three to five mounting screws holding the record changer to the cabinet from the bottom. Remove the record changer from the bottom. Remove the star wheel (62) from the bottom of the spindle. Earlier models have blade screws that are easily removed. Later models used recessed Phillips screws that can only be removed with the correct fitting screwdriver. If the star wheel is still stuck to the shaft after the screws are loosened, apply some penetrating oil. DO NOT FORCE BY PRYING. THE STAR WHEEL IS VERY BRITTLE AND WILL BREAK EASILY. Once the star wheel is removed, remove the two "C" clips (60, 20) that are holding the turntable shaft in place. Slide the spindle out from the top of the record changer.

The rubber idler wheel will now be visible on the top side of the record changer. Remove the idler wheel by either prying a spring or clip off the top of the shaft (if so equipped) or by removing the nut at the bottom of the idler shaft. Remove three nuts from the top of changer that are holding the motor in place. Note how all the washers and sleeves are assembled along with the rubber bushing. The bushing may well be so crumbled so that most of it is missing. You will be installing new bushings so throw out the old and clean off the sleeves because the old bushing material tends to stick to the metal parts. This can be accomplished with your fingernails or a flat bladed screwdriver. Slide the motor out of the bottom of the changer. If the rubber idler assembly comes out with the motor, remove this top assembly from the motor. If the motor spins freely it may only need a good oiling before being put back into service. Lift the top of the motor shaft to get the armature in its highest position. Apply oil above the two bearing caps along the shaft. Spin the shaft a few times and see if it is free. If it is binding, refer to the section on motor restore.

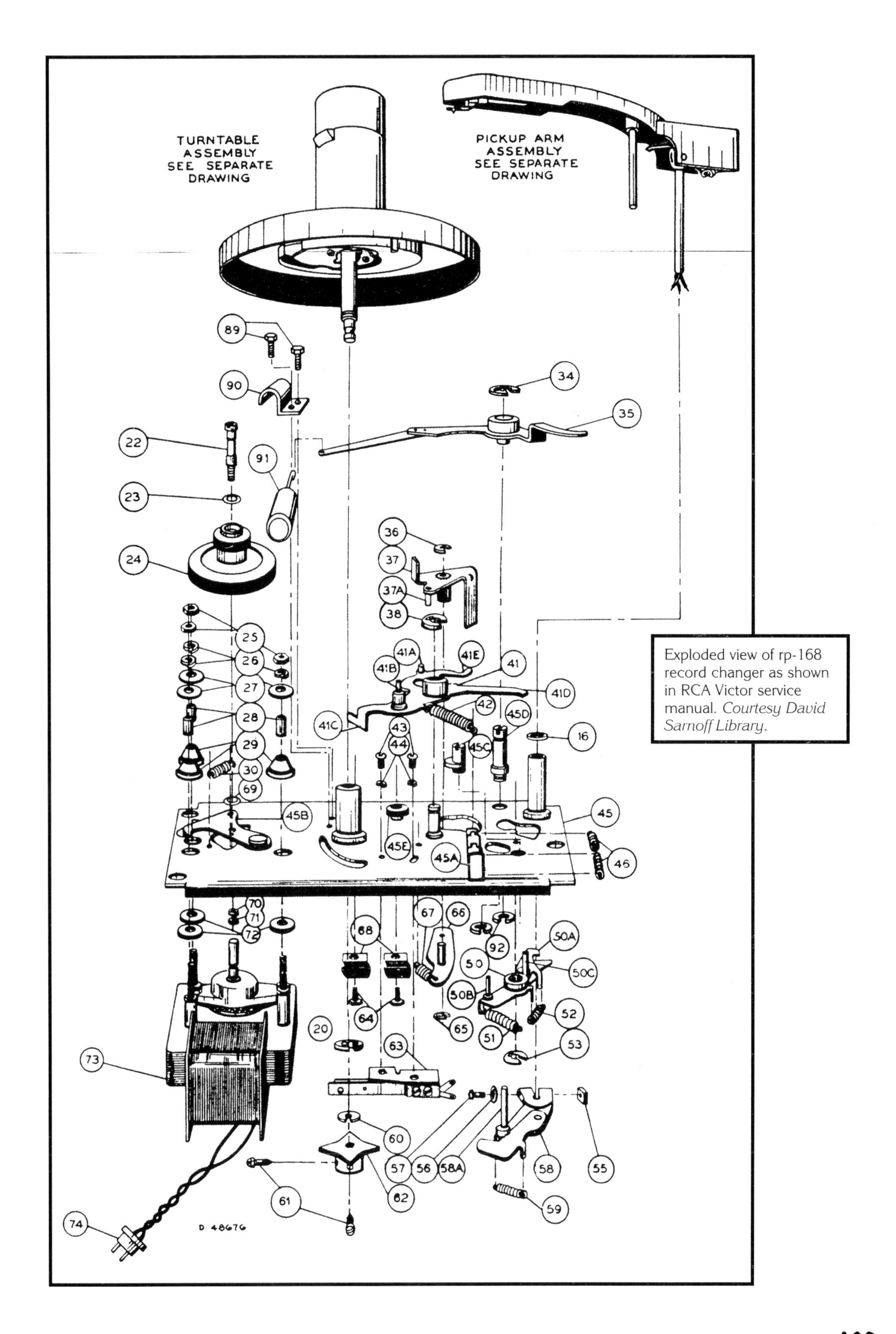

Exploded view of rp-168 record changer as shown in RCA Victor service manual. *Courtesy David Sarnoff Library*.

Motor Restore

Note: Some of the motors manufactured by Alliance for the Canadian market are made with pot metal bearing caps. This metal is very weak and usually falls apart with the slightest force applied. I do not recommend taking these motors apart. If oiling the bearings does not free up the motor, I would replace it. These motors will be stamped with "Alliance, made in Canada."

If the motor is seized up or doesn't turn free when you give it a spin, it will need to be taken apart and cleaned. Remove the two nuts and bolts that are next to the armature and holding the bearing end caps together. Now gently pull apart the end caps. You may need to use some penetrating oil to get the components apart. DO NOT FORCE because you may ruin the bearing cap. Clean the end shafts of the armature with alcohol. Use pipe cleaners to clean out the old grease from the end caps. Put one drop of light machine oil on the end shaft and insert it into the bottom end cap. If the shaft does not want to stay all the way in and floats on top of the oil, then remove the shaft, take some of the oil out and try again. DON'T ASSEMBLE WITH THE SHAFT FLOATING BECAUSE THE MOTOR WILL NOT RUN PROPERLY. Oil the top shaft close to the armature where the top bearing will sit. Now slide the armature through the windings, slide on the top bearing and snug up the two nuts and bolts with your fingers. DO NOT TIGHTEN. Rap the side of the coil windings with a screwdriver handle a couple of times. Then spin the shaft to see that it is free. Tighten the two nuts and bolts a little. Rap the coil windings again and spin the shaft. Do this several times until the nuts and bolts are adequately tight and the armature shaft spins freely. Rapping the side of the coil windings is essential to line up the bearings with the armature shaft. Run the motor and gently touch the spinning shaft with some emery cloth until the shaft shines. DO NOT OVERDO THIS TREATMENT. The idea is to remove any rust and dirt but not any steel. Doing so would change the speed of the record changer. Finish off by cleaning the shaft with alcohol.

There is a spring mounted on the motor that keeps the idler against the motor shaft. This return spring can lose some of its "spring" over the years. The combination of hardened idler rubber and loss of spring tension can cause the record changer to stall during the reject cycle, which produces the maximum load condition. If you can't replace the spring, cut off the last three windings and reshape the end in a loop. This will improve the return spring tension. It is not wise to increase tension too much because it will promote flat spots on the idler wheel when the unit is not in use.

Phonograph motor showing mounting screws.

At this point we will be removing all the levers that are held in place with "C" clips on the top and bottom of the changer. Remove the "C" clip that holds the director lever (41C) on the top of the changer and remove it. Remove the "C" clip that holds the pickup arm lift lever (35) and remove it. Remove "C" clip that holds the trip pawl (37) and remove it. Remove "C" clip that holds the trip pawl lever (66) and remove it. Remove the pneumatic dashpot (91) if so equipped. Clean the sub base (45) of all dust, dirt, oil, and grease with alcohol. Clean the levers and any shafts in the levers. Lightly oil all shafts and pivot points with light machine oil. With alcohol, clean the inside of the pneumatic dashpot housing and piston, if so equipped. Make sure parts are dry before reassembling. DO NOT LUBRICATE THE PNEUMATIC DASHPOT. The correct action is caused by moving air past the piston. If it is lubricated, the air will have a tougher time getting past the piston retarding the action. If the pneumatic dashpot is still sluggish it may require honing of the inside of the housing with emery cloth. Simply pull the piston all the way out of its cylinder. Wrap emery cloth around a pencil shaft and insert it in the cylinder. Stroke the inside evenly around. Clean again with alcohol. Reassemble the cleaned levers and pawls with their "C" clips. Reassemble the pneumatic dashpot. Temporarily insert the tonearm back in the sub base (45) and push the pickup arm lift lever (35) so that the tonearm lifts up. Then let go of the lever and watch the tonearm descend. This will be how fast the tonearm will descend onto the record via the pneumatic dashpot. If the action does not appear to be correct, this is the time to go back and work with the pneumatic piston. This action is also affected by the weight of the tonearm, so if you are going to use a different cartridge, install it before you do this test (refer to Cartridge Replacement, above).When you're done with this test, remove the tonearm for now. If idler wheel is worn or dried out, replace it with a new one (easier said than done). An idler wheel cost eighty cents back in the 1950s. Now they are close to impossible to find and companies are charging up to thirty dollars to rebuild them. Check to see if the existing idler is still round. Do this by applying power to the motor with the idler installed. Listen for any thumping noises or observe any vibrations from the idler wheel. If it appears round but is hard and glazed, you can improve its grip with the procedure described next.

Idler Wheel Refurbish

Clean the outer surfaces with lacquer thinner. CAUTION! Lacquer thinner can dissolve rubber so follow these instructions carefully. Put some lacquer thinner on a soft clean cloth and wipe the outer surfaces of the wheel. You will notice plenty of black coming off as it cuts through the outer glazed rubber. Immediately after applying the lacquer thinner, remove it with a clean cloth. Do this two or three times to get rid of all the glazed rubber. You will find the edges of the idler will now be much more resilient and have more grip. There are other products on the market that improve grip but they usually last only a short time. These include applying rubber revitalizer available from Projector Recorder Belt Company or phono no slip compound (rosin) available from GC Electronics. Even the lacquer thinner treatment doesn't last forever but lasts longer than the other treatments.

Clean and lubricate the return lever (45B) so that it moves freely in all directions.

If the motor spins freely it may only need a good oiling before being put back into service. Lift the top of the motor shaft to get the armature in its highest position. Apply light machine oil above the two bearing caps along the shaft. Spin the shaft a few times and see if it is free. If it is binding, refer to the section on motor restore. Install three new rubber grommets (29) on the motor mountings and reinstall motor on sub base. The cone end of the grommets should be on the top of the sub base. If it is installed backwards the motor will sit too low. Install another three grommets on the outer edge of the sub base where the changer meets up with the mounting holes in the Bakelite cabinet or metal motorboard. These grommets should have the cone facing downwards. Some early changers are mounted with springs instead of rubber grommets. In this case, reinstall the springs.

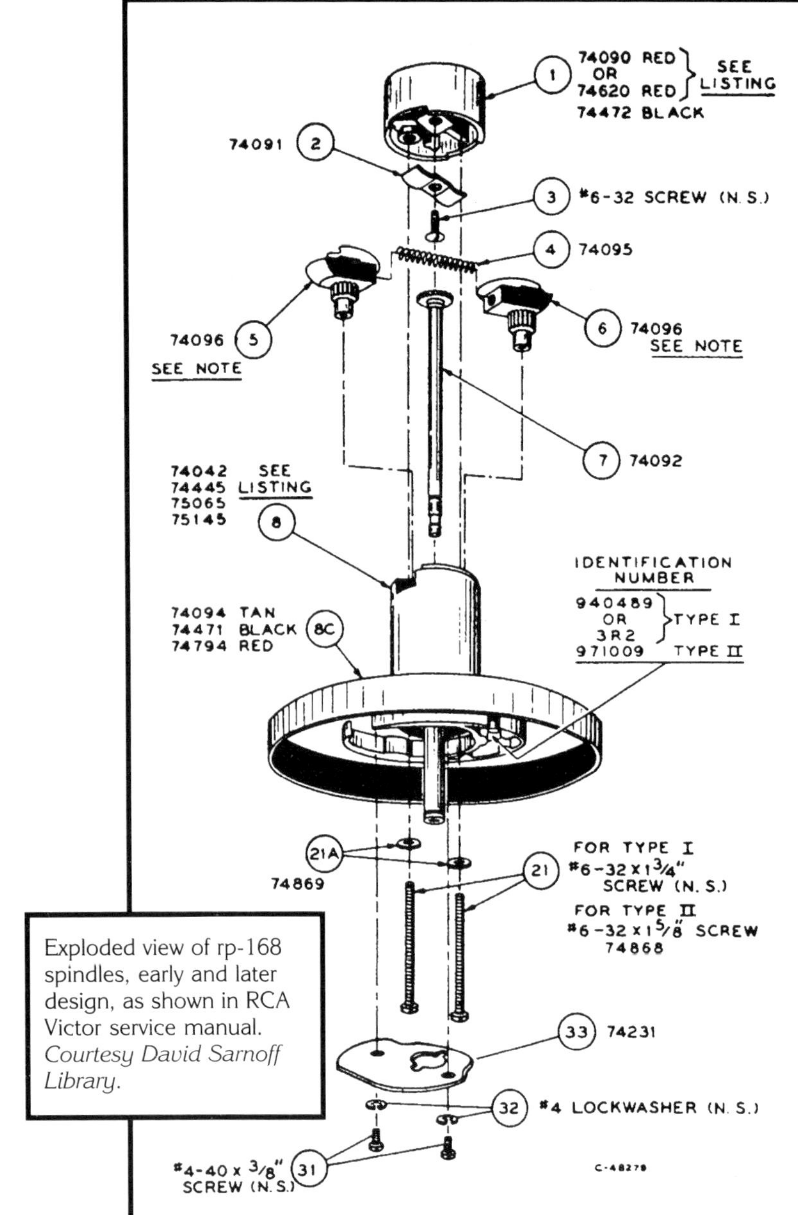

Exploded view of rp-168 spindles, early and later design, as shown in RCA Victor service manual. *Courtesy David Sarnoff Library.*

Figure 20—Turntable Assemblies, Types I and II.

Main Lever vs Record Separators:

Two different main levers (director lever) are used depending upon the type of record separators being used.

Stock No. 74076 lever is used only with the rotating gear type of record separators. The end (41C) that engages the star wheel is long.

Stock No. 74857 lever is used only with the push-out type of record separators. The end (41C) that engages the star wheel is short.

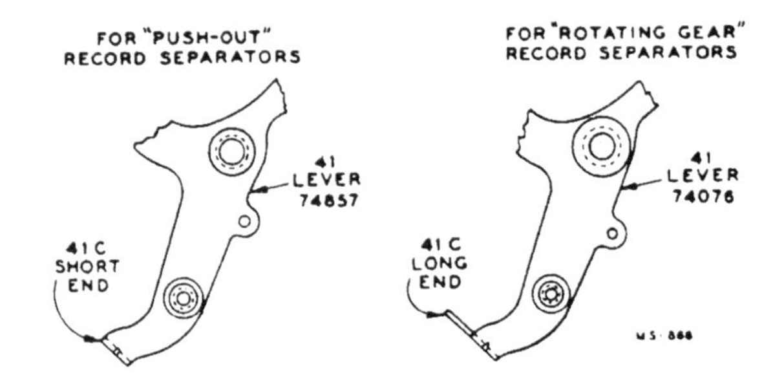

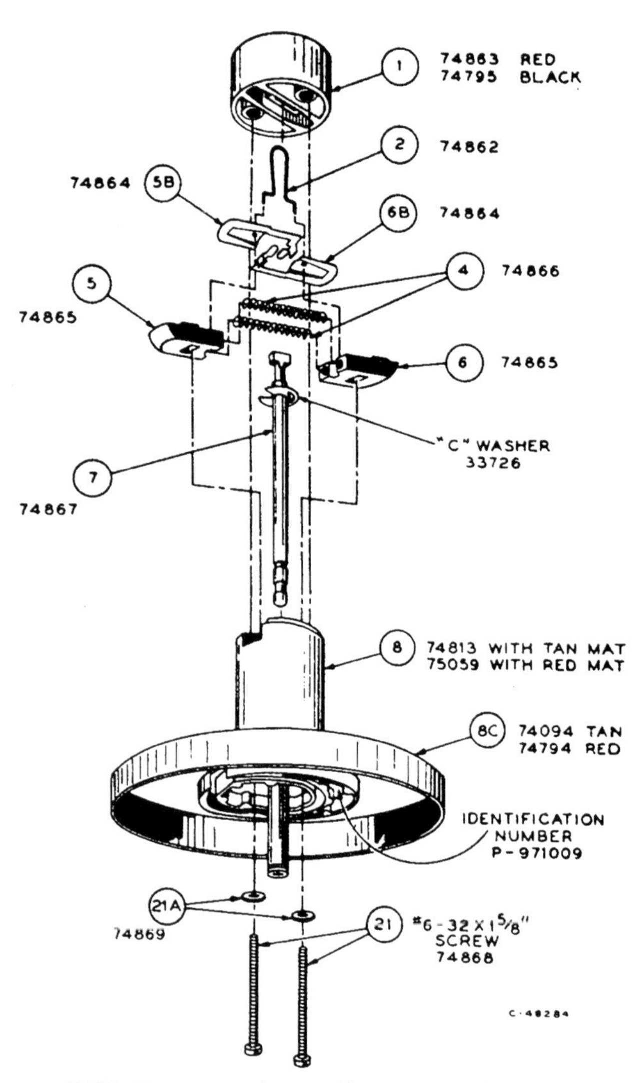

NOTE: Use care in dis-assembly to prevent loss of springs. Remove screws—lift nose slightly—hold both separator knives down against shelves—then remove nose.

Figure 21—Turntable Assemblies, Type III.

TURNTABLE ASSEMBLIES

Type I

Turntable Stock No. 74042. Stamped 940489 or 3R2. Has TAN MARBLEIZED mat and uses rotating gear type of record separators. Use No. 74090 spindle nose—RED (thin wall)

Turntable Stock No. 75065. Same as No. 74042, except for diameter at top of spindle. Use No. 74620 spindle nose—RED (thick wall)

Turntable Stock No. 75145. Same as No. 75065, except that it has a RED mat. Use No. 74472 spindle nose—BLACK

Turntable Stock No. 74445. Same as No. 75065, except for finish and BLACK mat. Used only on Model CP-5203. Use No. 74472 spindle nose (BLACK)

Type II

Stamped 971009. Follower cam (33) is a part of the turntable casting. Otherwise, similar to No. 75065. Use No. 75065 turntable, and No. 74231 cam for replacement

Type III

Stock No. 74813. Stamped 971009. Has TAN MARBLEIZED mat and uses push-out type of record separators. Use No. 74863 spindle nose—RED. Although this turntable bears the same stamping as Type II, it does not have the shafts required for mounting the rotating gear type of separators

Stock No. 75059. Same as No. 74813, except that a RED mat is used. Use No. 74795 spindle nose—BLACK

NOTE: Main Lever (41)

Stock No. 74076 lever (with long end 41C) is used in conjunction with rotating gear type of record separators. Stock No. 74857 lever (with short end 41C) is used in conjunction with push-out type of record separators

Rp-168 Spindle Rebuilding

An explanation is needed to understand the two different methods of dropping the records on the spindle. The early design depends on gear rotation to move the shelves and separators. Think of the separators as little cams so that as they rotate, the amount of separator sticking out of the spindle will change, providing the separating motion. The later design changes the rotating motion to in/out motion at the center of the spindle where the center shaft meets the shelves and separators.

Turn the spindle upside down and remove the two screws that are holding the cam follower plate near the center of the spindle. Later spindles do not have a cam follower plate. The same action is accomplished by changes in the casting. Remove two screws that are recessed in the center while you are holding the spindle cap in place. Once these screws are removed, the spindle cap and all its parts will come off. Carefully remove them. It is a good idea to first study the diagrams to see how the parts fit together because once they come apart, it may appear confusing. Some of the parts are spring loaded, so keep one hand around them while prying them apart with the other hand. Once the shelves, springs, and separators are removed, take the center shaft out through the top of the spindle. Clean any dust, dirt, and grease off these parts. The only part that should be lubricated with light machine oil is the center shaft. Putting lubrication on the shelves and separators will cause dust to collect quickly and affect the assembly's ability to drop the records. Use a pipe cleaner and some alcohol to clean the shaft in the center of the spindle. Relube the center shaft and place it back in the spindle. Replace the shelves, separators, and springs on the top of the spindle as follows:

Early Design (Rotating Separators)

For the early type rp-168 with rotating separators (shown on left in exploded view), the separator (5) consists of a shelf, separator, and gear assembly. Place one of them in the spindle and mesh the gear with the center shaft gear. Place the return spring (4) into the side of the mounted separator (5). Now put the other end of the return spring (4) into the other separator (6) assembly, push it into the spindle, and mesh it with the center shaft gear. It is crucial that the orientation of the separators is the same. If they are not quite the same, lift one of the separator assemblies (5) and move it over by one gear tooth so that the orientation is the same. While you are doing this use your other hand to keep the return spring (4) from popping out and disappearing somewhere in a corner. Place the spindle cap (1) on top of the spindle and line it up so that the separators are under the notches on the spindle cap (1). Install the two screws that hold the cap to the spindle. CAUTION! The plastic spindle cap is quite brittle and you must avoid the temptation to really tighten the screws. Just snug the screws. Twist the bottom of the center shaft (7) and observe the motion of the shelves and separators rotating. There should be some minor resistance as you go through the motion due to parts working against the return spring. A good test at this point is to put a stack of records on the top of the spindle and rotate the center shaft and watch the records fall one by one. If the records don't fall properly open the spindle again and find out what you did wrong. Either the separators are not aligned properly or the return spring is weak.

Later Design (Push In/Push Out)

For the later type rp-168 with push out separators (shown on right), place the two return springs (4) in between the two shelves (5, 6) and place it all into the top of the spindle. Place the two separators (5b) and (6b) on top of one another as shown in the exploded view. Now place them on top of the shelves and springs. With your index fingers, push the shelves into the spindle until the separators fall into place. This might take a little practice so take your time and try to get the feel for this operation. Once these components are in place, connect the return spring (2) on top of the separators. You will have to open up the spring in order to get it to enter the proper holes on top of the separators. When properly installed the return spring will stand straight up in the center of the spindle. Use care during these operations because if the parts are not correctly lined up, they could come apart with great velocity and never be seen again. Place the spindle cap (1) on top of the spindle by lining up the return spring so it goes in the slot in the middle of the spindle cap. Install the two screws to hold the spindle cap to the spindle. CAUTION! The plastic spindle cap is quite brittle and you must avoid the temptation to really tighten the screws. Just snug the screws. Twist the bottom of the center shaft (7) and observe the motion of the shelves and separators. Some minimal resistance should be felt when the shelves go in and the separators come out. A good test at this point is to put a full stack of records on the top of the spindle and rotate the center shaft and watch the records fall one by one. After the first record falls, make sure the rest of the stack drops down onto the shelves. If the stack sometimes stays on the separators and doesn't drop onto the shelves, it means the return spring (2) is too weak and needs to be replaced. Since the return spring is not available from RCA Victor anymore, I have resorted to using two return springs together. Another solution is to make a home made spring from a bobby pin. Cut the bobby pin to the correct length.

Clean the inside of the turntable rim with alcohol and clean and lube the cam slot underneath the spindle where the director lever will follow the cam. The spindle is now rebuilt and ready for service.

Spindle caps, early and later design.

Rp-168 Reassembly

Clean and lube the turntable thrust bearing and two washers that go on the turntable shaft. It is extremely important that the washers are absolutely free of any dirt or grit, because any such foreign substance will cause the turntable to flutter. Remove the old grease in the bearing and relube. Note: The exploded view of the rp-168 does not show the turntable thrust bearing and the two thrust washers. They are installed on the shaft protruding from the bottom of the turntable. First slide on a washer, then the bearing, and then another washer. Place the spindle into the center hole on the sub base and rotate it as you gently push it all the way in. The rotation will allow you to get the spindle rim past the idler wheel. DO NOT FORCE THE SPINDLE IN PLACE. Support the record changer by the ends of the sub base so that it can be plugged in and observed. Once power is applied the turntable should turn freely. While the turntable is rotating, gently grasp the top of the spindle and see how much force is required to stop the rotation. There should be a fair amount of resistance to stopping the rotation. If it stops too easily, it means that the idler wheel does not have a sufficient "grip" on the inside of the turntable rim. If the idler wheel has been reconditioned as completely as possible, and it still doesn't have adequate "grip," the fault may be that the return spring (43) is stretched, and is no longer sufficiently taut. Cutting off about three curls at one end and reshaping the end of the spring into a loop may correct this condition. Don't cut off too much spring, because the idler wheel will then push too hard against the motor shaft, causing erratic speed and flat spots on the idler wheel. Push the reject button and see if the turntable rotates during the reject cycle. Install the tonearm on the changer. Make sure that the fiber washer is still on the shaft of the tonearm. Note: If you are replacing the cartridge, it should have been done while the tonearm was removed. Refer to cartridge replacement section. Poke the tonearm wires through the clamp and trip lever (58) and slide the clamp and trip lever onto the tonearm shaft. At this point the following adjustments need to be made (refer to Appendix A):

A. Tonearm alignment
B. Director lever position
C. Landing adjustment
D. Pickup height adjustment

Once adjustments are satisfactorily made, re-solder the cartridge wires. Make sure they are located so they do not get tangled in the cycling mechanism. This also goes for other wires that are under the set.

Note: I find that once the changer is back together it is very helpful to play several stacks of records and observe the operation. It will not be unusual to go back and tweak some of the adjustments.

Rp-190 Record Changer

The rp-190 changer was released sometime in 1950 and was produced from 1950 through 1958. This record changer is much more modern looking than the rp-168, sporting a plastic turntable and tonearm. The reject mechanism was completely redesigned to be smoother in operation. No longer tied to only one revolution of the turntable, the reject is accomplished with a rubber cycling cam. The machine could automatically play up to fourteen records, but once again there was no automatic shutoff. The light plastic turntable required a perfectly circular rubber idler wheel, otherwise there was sufficient flutter in the turntable's operation to be irritating to the listener.

Metal, rather than plastic, turntables were used mostly on children's players; occasionally a metal turntable can be found on other players that incorporated the rp-190 changer.

The tracking problem described on the earlier rp-168 changer is evident on the rp-190 as well, and the "coin-weight" solution is often encountered when one has found an old 45 player for restoration. Trying to remove the thirty-year-old tape and coin from the soft plastic tonearm requires special attention, as the tape's adhesive has frequently bonded with the tonearm's plastic.

As the record changers age, the rp-190 is more prone to reject problems because it uses the rubber cycling cam, which gets hard and brittle.

Improvements Made in Late 1950

As shown on the right side of the exploded view, a spring and lever were added to the cycling slide (82) and two springs were removed at the back of the tonearm. These arrangements minimize erratic tonearm movement and jamming during the reject cycle.

The tonearm hold-down clamp was improved with a new type of clamping apparatus, which made it much easier to remove and install the tonearm.

Cost-Cutting Measures Introduced in 1955

The mute switch was replaced with a simple terminal strip. The cycling cam thickness was reduced, thereby requiring less rubber per cam. The extra spring and lever on the cycling slide that were added in 1950 were removed.

Disassembly

From the cabinet, remove the bottom panel, if so equipped. Remove tonearm hold-down bolt (33), and pull tonearm up and out. Remove the cartridge, with its clips and wires, from the tonearm. The plastic tonearm is very soft plastic and care must be taken not to damage the finish. Also, do not break the plastic pins that hold the clips.

Sometimes it is difficult to remove the wires from the rear clip (26). If so, you will need to remove the counterbalance and swivel assembly (28). Remove screw (30) and gently pry at the assembly because it is glued in place. If you are not careful you can crack the tonearm. All components are now removed from the tonearm. From underneath the motorboard (9), remove the 3 "C" clips (47) holding the motor in place. Now remove the washers (46). Pull the motor out from the bottom. Remove the idler wheel (39) by prying the spring (37) from the top of the idler shaft. In order to separate the motor and amplifier (if so equipped) remove the screw (61) holding the on/off switch, and remove the switch. Remove also the screws (67) holding the muting switch (68) or terminal strip (cost reduction) and remove it from the changer. Now the changer is free from the other components and will be easier

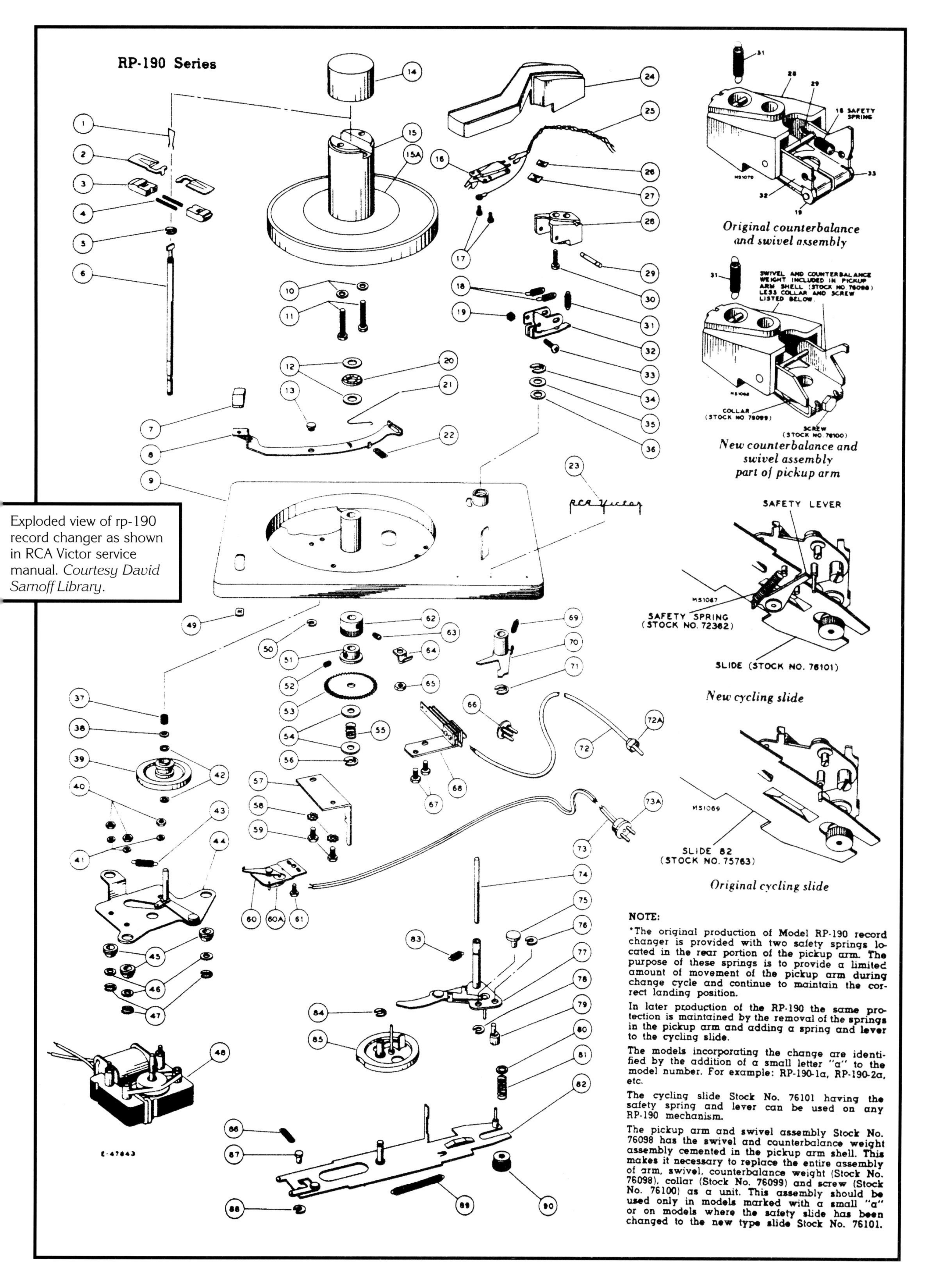

Exploded view of rp-190 record changer as shown in RCA Victor service manual. *Courtesy David Sarnoff Library*.

to work on. From underneath the changer remove spring (89) and "C" clip (56) from the center of the spindle. This "C" clip is spring loaded, so keep your fingers on the clip as you are sliding it out. Remove washers (54) and spring (55). Now you can remove ratchet wheel (53). Use an Allen wrench to remove the friction collar (51). Next use a flat-bladed screwdriver to loosen the screw on the knurled roller (62). Pull the spindle from the top and it should slide out. If it still does not want to slide out, put some penetrating oil on the center spindle shaft to loosen the knurled roller. Next unscrew the knurled nut (90) until it comes completely off. Lift the cycling slide (82) and gently slide it out from the mounting bracket (57). It may hang up at the pin on the cycling cam (85), so observe this possibility as you slide it out. Remove "C" clip (84) and remove the cycling cam from the slide. Remove "C" clip (34), washer (35), and spacer (36) from the top of the changer. The trip lever assembly (77) should now slide out of the bottom of the changer. Remove spring (81), washer (80), and "C" clip (71) from bottom of changer. Remove spring (69) and pull off the return lever (70). Remove RCA Victor logo (23) by removing pin retainers or wax that is used to hold the logo in place. This is enough disassembly to do a thorough overhaul. It is possible to remove the remaining parts, but after years of experience, I have found that some of them are difficult to remove, *e.g.*, the reject lever mounting stud (13); in addition, some parts are difficult to put back on, *e.g.*, the control arm speed nut (49). Just lightly lube the pivot points.

Apply power to the motor with the idler installed. Listen for any thumping noises or observe any vibrations from the idler wheel. If it appears round but is hard and glazed, you can improve its grip by following the procedure described earlier in the section on "Idler Wheel Refurbish."

On reassembly, clean and lubricate the return lever on the motor mounting (44) so that it moves freely in all directions. Install three new motor bushings (45) and reinstall the idler wheel and spring cap (37). It is important that the spring be pushed firmly on, so that there is minimal vertical movement of the idler wheel. If the motor spins freely it may need only a good oiling before being put back into service. Lift the top of the motor shaft to get the armature into its highest position. Apply light machine oil above the two bearing caps along the shaft. Spin the shaft a few times and see if it is free. If it is binding, refer to the earlier section on "Motor Restore." The idler wheel tensioning spring (43) provides the correct "grip" between the motor shaft, the idler wheel, and the rim of the turntable. If slippage occurs after the rebuild, it may be necessary to replace or shorten this spring (43). To shorten, cut three windings from one end and bend the shortened end into a loop.

Rp-190 Spindle Rebuilding

Turn the spindle upside down and remove the two screws (11) and washers (10) that are recessed in the center, while you are holding the spindle cap in place. Once these screws are removed, the spindle cap (14) and all of its parts will come off. Carefully remove them. Some of the parts are spring loaded, so keep one hand around them while taking them apart with the other hand. Once the shelves, springs, and separators are removed, take the center shaft out through the top of the spindle. Clean any dust, dirt, and grease off these parts. The only part that should be lightly oiled is the center shaft. Putting lubrication on the shelves and separators will cause dust to collect quickly and affect the assembly's ability to drop the records correctly. Use a pipe cleaner and some alcohol to clean the shaft in the center of the spindle. Lubricate the center shaft with light machine oil and place it back into the spindle. Replace the shelves, separators, and springs on the top of the spindle as follows:

Put the two return springs (4) in between the two shelves (3) and place it all into the top of the spindle. Place the two separators (2) on top of the shelves and springs. With your index fingers, push the shelves into the spindle until the separators fall into place. This might take a little practice, so take your time and try to get a feel for this operation. Once these components are in place, connect the spindle cap spring (1) on top of the separators. You will have to open up the spring in order to get it to enter the proper holes on top of the separators. When properly installed, the return spring (1) will stand straight up in the center of the spindle. Use care during these operations, because if the parts are not correctly lined up, they could come apart and land across the room. Place the spindle cap (14) on top of the spindle by lining up the return spring so it goes into the slot in the middle of the spindle cap. Install the two screws to hold the spindle cap to the spindle. CAUTION! The plastic spindle cap is quite brittle and you must avoid the temptation to really tighten the screws. Just snug the screws. Twist the bottom of the center shaft (6) and observe the motion of the shelves and separators. Some minimal resistance should be felt when the shelves go in and the separators come out. A good test at this point is to put a full stack of records on the top of the spindle, rotate the center shaft, and watch the records fall one by one. After the first record falls, make sure the rest of the stack drops down onto the shelves. If the stack sometimes stays on the separators and doesn't drop onto the shelves, it means that the return spring (1) is too weak and must be replaced. Since RCA Victor can no longer supply the return springs, I have resorted to using two used return springs together. Another way to handle this is to cut a bobby pin to the correct length and use it in place of the original spring (1). Finish the job by cleaning the inside of the turntable rim with alcohol. The spindle is now rebuilt and ready for service.

Cycling Cam Refurbishing

After trying many different ways of improving the "grip" of the cycling cam, I have determined that the following is the most dependable: Clean the old rubber with lacquer thinner to remove any dirt, grease, or loose rubber. Make sure you remove all traces of the cleaner with a clean dry cloth. Pour some "Plastic Dip," a coating that goes on wet and dries looking like rubber, into a small container. Hold the cam vertically, so that the bottom part enters the liquid. Move the cam steadily so that the outer rubber gets coated with a thin coat of the Plastic Dip. Let the cam dry overnight. After the liquid completely dries it forms a very thin and durable rubber coating that has plenty of "grip." Usually the coating is thin enough so that it doesn't affect the record dropping adjustment, but if two records tend to fall instead of one, refer to "record dropping adjustment" in Appendix A.

Another good way to refurbish the cycling cam is to remove the old rubber completely and install new rubber that can be purchased (refer to Appendix C).

Refurbishing old cycling cam with "Plastic Dip."

Rp-190 Reassembly

Refer to Service Tips (page 148) for information on removing any residue from the plastic tonearm. Reinstall the return lever (70) and spring (69). Install the clip (71), washer (80), and spring (81). Push the return lever (70), and see that it springs back to the returned position. Install the trip lever assembly (77) through the bottom of the changer base and install spacer (36), washer (35), and clip (34) from topside of the base. Make sure that the shiny side of the clip is facing downward.

Reinstall a new or refurbished cycling cam on the cycling slide (82) with a liberal amount of grease. This tends to stop the cam when it reaches the end of the cycle and enters the notched-out rubber portion of the cam. Reinstall the cycling slide (82) by sliding it into the notches in the bracket (57). The pin sticking out of the cycling cam (85) must go through the slot in the motorboard (9). The other end of the slide should rest on the return shaft (70) and spring (81). To accomplish this, slide the height adjustment nut (90) into the slot at the end of the cycling slide, and align the nut with the threads on the return lever (70). Because the spring (81) will produce resistance, you will have to push the nut to start it on the threads. Screw the nut (90) clockwise part of the way down. This nut will be adjusted later, to achieve the correct tonearm height during the reject cycle.

Clean and lubricate the turntable thrust bearing (20) and the washers (12) that go between the turntable shaft and the sub base. It is extremely important that the washers are absolutely free of any dirt or grit, because any foreign substance will cause the turntable to flutter. Remove the old grease in the bearing and lubricate it with light grease. Place the spindle into the center hole on the sub base and rotate it as you gently push it all the way in. The rotation will allow you to get the spindle rim past the idler wheel. DO NOT FORCE THE SPINDLE INTO PLACE.

Support the record changer by the ends of the motorboard so that it can be plugged in and observed. Once A.C. power is applied, the turntable should turn freely. As the turntable is rotating, gently grasp the top of the spindle and see how much force is required to stop the rotation. There should be a fair amount of resistance to stopping the rotation. If it stops too easily, it means that the idler wheel does not have a sufficient "grip" on the inside of the turntable rim. If the idler wheel has been reconditioned as completely as possible and still does not have adequate "grip," the fault may be that the return spring (43) is stretched and is no longer sufficiently taut. Cutting off about three curls at one end and reshaping the end of the spring into a loop may correct this condition. Don't cut off too much spring, because the idler wheel will then push too hard against the motor shaft, causing erratic speed and flat spots on the idler wheel. Note: If you are replacing the cartridge, it should have been done while the tonearm was removed. Refer to cartridge replacement section. Install cartridge wires and retaining clips (26, 27). Poke the tonearm wires through the slot on the motorboard near where the tonearm mounts. Lightly snug up the clamp by tightening screw (33).

At this point you can put a single record on the spindle manually and play the machine through an amplifier to see how it sounds and to observe the action of the reject cycle. It's easier to stop here and go back if something isn't right. Once you are happy with the performance of the changer, you can finish the assembly. Pull the spindle back out by grabbing the top and pulling straight up. Note where the hole is in the shaft (6). This is where the knurled roller (62) screw should enter the turntable shaft. Push the bottom of the turntable shaft through the motorboard center hole until it just starts to come out of the bottom side of the motorboard. Slide the knurled roller (62) behind the cycling slide and let the spindle shaft (6) go through the knurled roller. Now line up the hole in the spindle shaft (6) with the setscrew in the knurled roller (62). Tighten the setscrew. This operation can be tedious, and some repair people do not take the time to find the hole in the turntable shaft. Instead, they tighten down the setscrew on the shaft itself. There is another shaft running through the larger shaft, and tightening the setscrew down on the outer shaft can cause binding between the inner and outer shafts. Next, slide the friction collar (51) onto the bottom of the spindle shaft (6) with the larger diameter facing away from the spindle. Gently pull on the bottom of the spindle shaft to make sure that it is fully extended, and then tighten the setscrew on the friction collar. Put the ratchet wheel (53) on the spindle shaft (6) with the letter "S" facing away from the spindle. The teeth along the edge of the ratchet wheel will not mesh correctly if the wheel is reversed. Install the washer (54), spring (55), and another washer (54). Now install clip (56). Finally, install the slide assembly return spring (89). Now you can try the reject cycle by pushing the reject button and turning the turntable clockwise.

At this point the following adjustments need to be made (refer to Appendix A):

A. Rough tonearm alignment
B. Landing adjustment
C. Pickup height adjustment

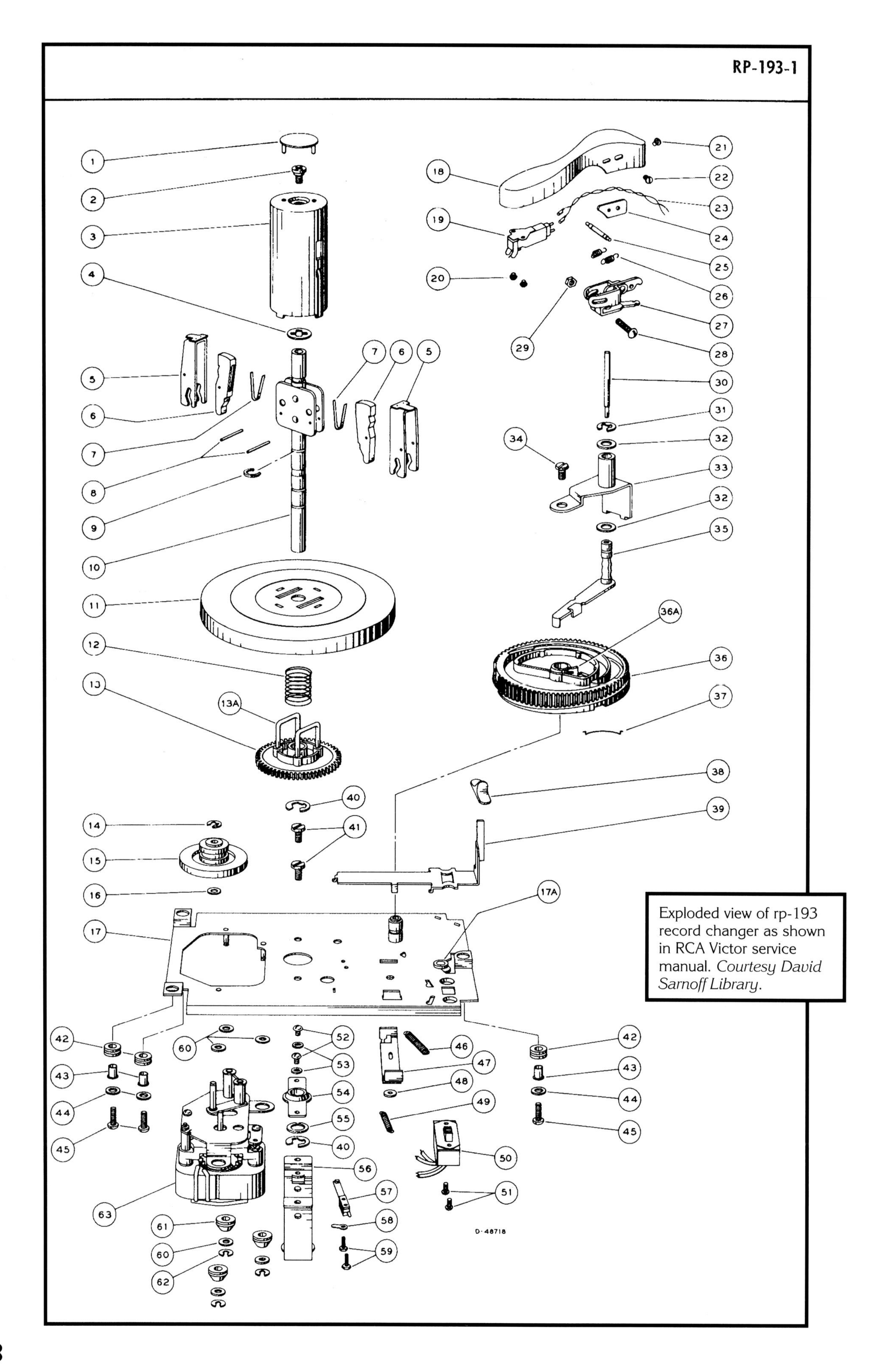

Exploded view of rp-193 record changer as shown in RCA Victor service manual. *Courtesy David Sarnoff Library.*

Rp-193 Record Changer

Disassembly

Remove the mounting screws from the bottom of the Bakelite case and remove the cardboard bottom. Disconnect the wires from the cartridge in the tonearm (18). Loosen the clamp screw (28) and lift the arm straight up and out. To remove the tonearm swivel bracket, remove the landing adjustment screw (22). Then push the pivot pin (25) away from the slotted side of the pickup arm shell, bend the shell slightly to remove the pin, and the entire swivel assembly will slide out. To remove the turntable from the changer, remove the "C" clip (40) below the upper turntable assembly. Pull the turntable straight up and out. To remove the cycling cam, remove screw (34). Lift the bracket assembly consisting of parts (31 through 35). Lift the cycling cam off the shaft.

A typical selenium rectifier.

Rp-193 Spindle Rebuilding

After removing the "C" clip (40), slide the pinion gear (13) and spring (12) off the spindle shaft (10). Remove the plastic cap (1). Remove screw (2). Separate the spindle cover (3) from the turntable. Push pin (8) out to remove spring (7), shelf (6), and blade assembly (5). Finally, remove "C" clip (9) and lift out the spindle shaft (10). After cleaning and careful reassembly, refer to appendix A for spindle adjustments.

Rebuild motor according to instruction in the "Motor Restore" section. The idler wheel is slightly different on this changer. I have successfully used idler wheels from rp-168 and rp-190 changers on the rp-193. The rubber wheel will grab a different section of the turntable rim but will still function adequately. The only difference between the idlers is the position of the sleeve that runs through the center of the idler.

Rp-193 Reassembly

Reinstall parts in reverse order. Make sure that the cycling cam turns freely once it is placed back on its shaft. If there is too much resistance here, the turntable will stall during the reject cycle. Make sure that the detent lever spring (17a) is lubricated with grease and runs smoothly on the cycling cam. Refer to Appendix A for turntable adjustments.

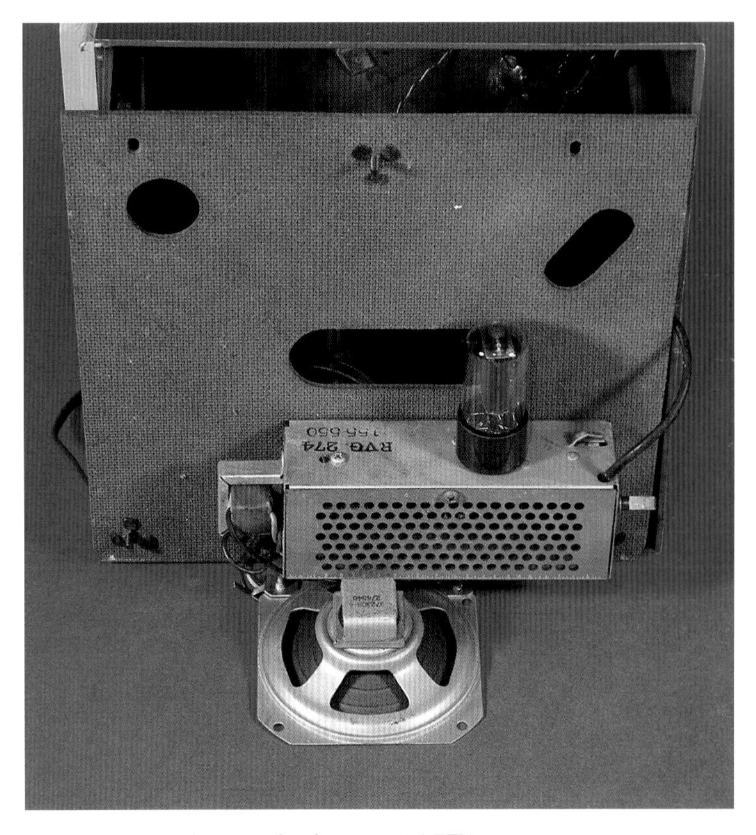

"One tube wonder" amplifier from model 7EY1.

Amplifiers

One-stage Audio with Single-Ended Output

This is the economical amplifier that RCA Victor used with the 45 phonographs. A selenium rectifier was used to provide the needed voltages for the single solitary audio stage. To cut costs, the amplifier was put into series with the phonograph motor, which was rated at 85 or 90 volts instead of 120 volts. In order to handle the anemic audio, a high-output cartridge was employed. The most commonly used tube was a 25L6. These amplifiers have been nicknamed "one tube wonders." Sound quality is mighty thin.

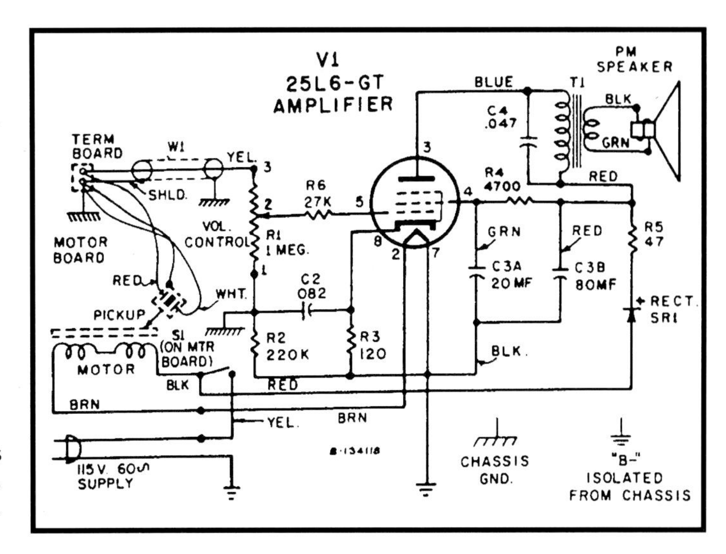

Schematic for "One tube wonder" as shown in RCA Victor service manual. *Courtesy David Sarnoff Library.*

Two-stage Audio with Single-Ended Output

Many of the early amplifiers had two audio stages, consisting of a 12AV6 preamplifier and 50C5 power amplifier. The rectifier was either a 35W4 tube or a selenium rectifier. The tube filaments were typically in series with a power resistor, adding up to 120 volts. Obviously these amplifiers have more volume than the single tube units. Sound quality is still on the tinny side, however, especially when the underpowered amplifier is driving the typical four-inch speaker.

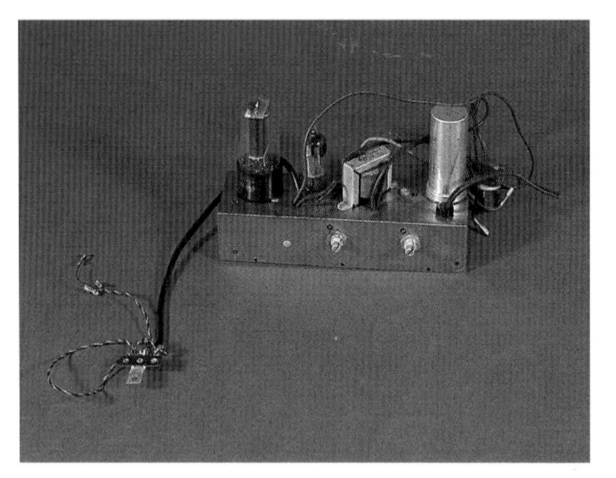

Amplifier from 7EY2 series.

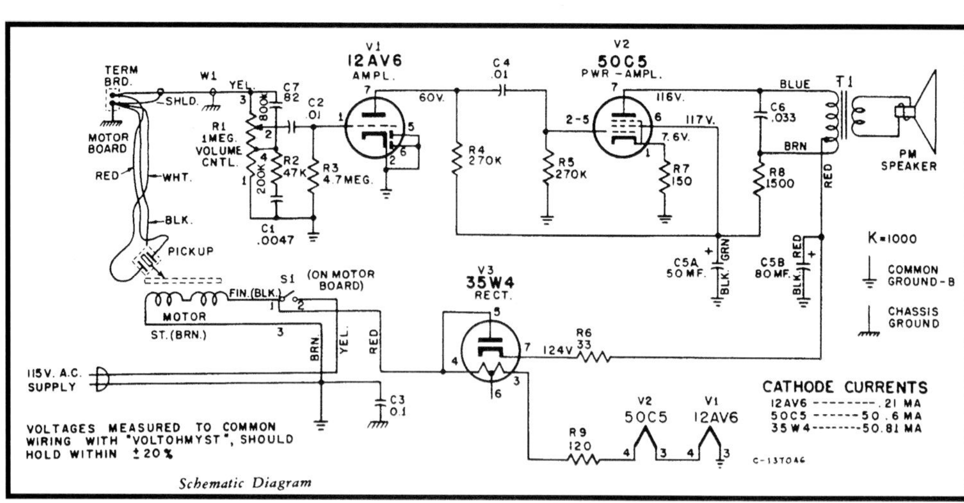

Schematic for two-stage audio amplifier model 7EY2 as shown in RCA Victor service manual. *Courtesy David Sarnoff Library.*

Model 9EY3 amplifier.

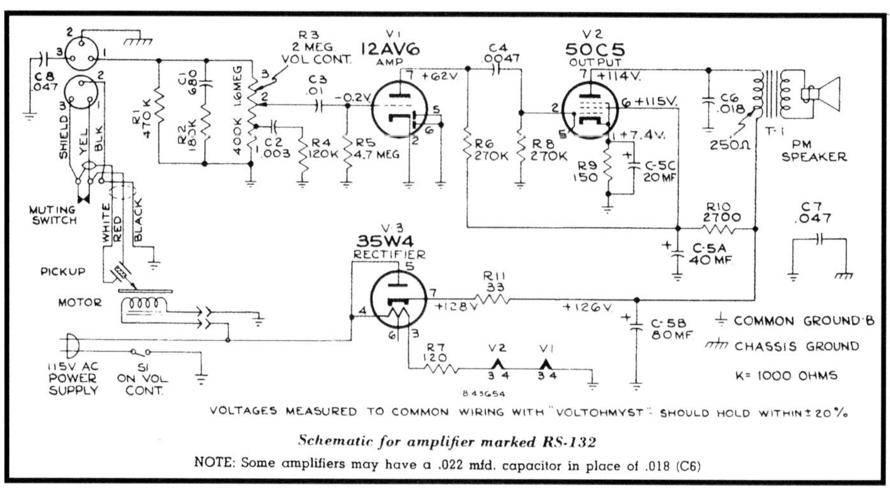

Schematic for two-stage audio amplifier model 9EY3 as shown in RCA Victor service manual. *Courtesy David Sarnoff Library.*

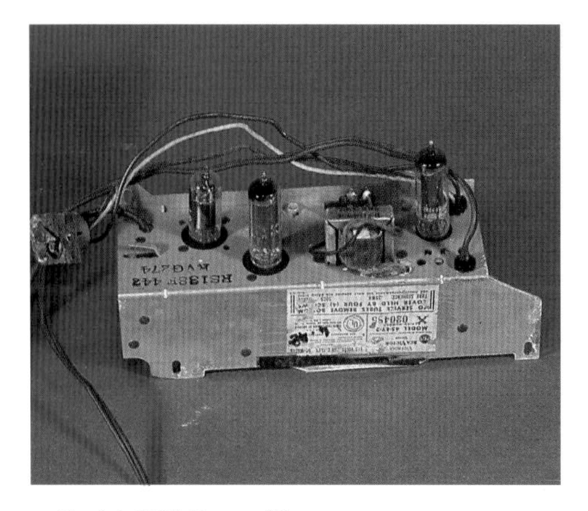

Model 45EY2 amplifier.

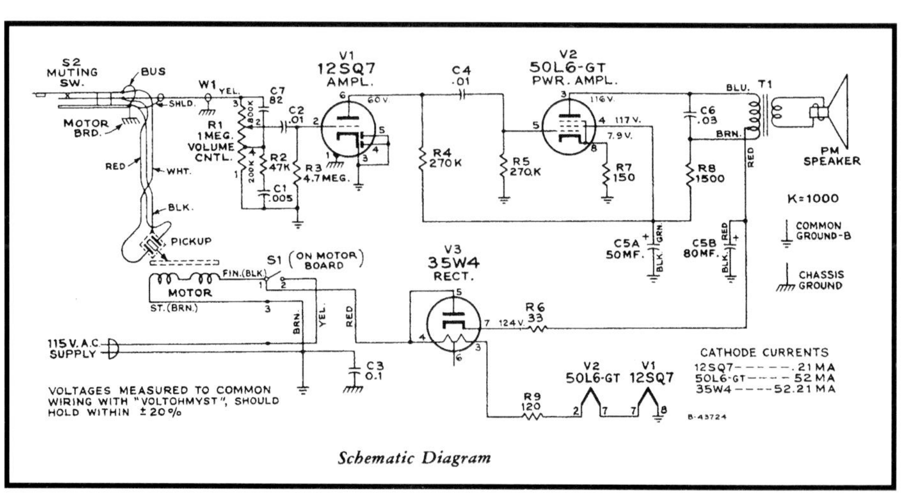

Representative schematic for two-stage audio amplifier model 45EY2 as shown in RCA Victor service manual. *Courtesy David Sarnoff Library.*

Model 45EY3 amplifier.

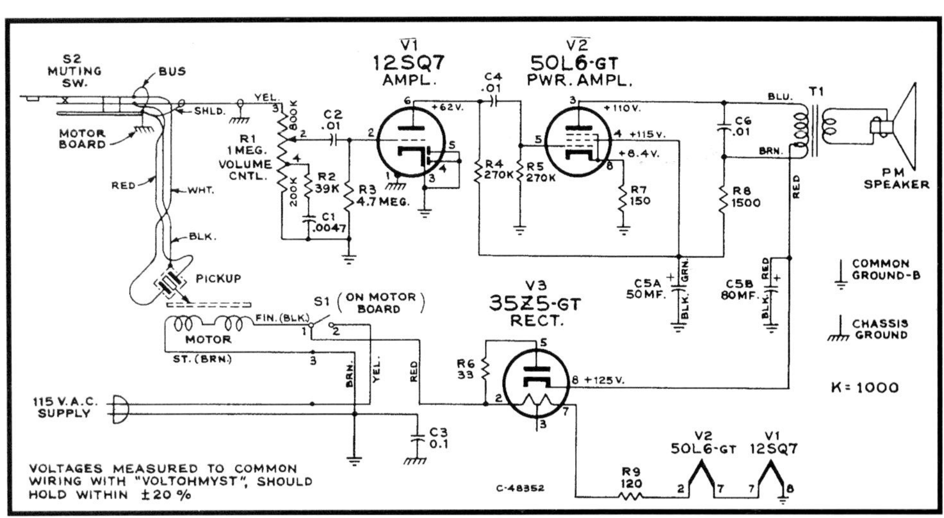

Representative schematic for two-stage audio amplifier model 45EY3 as shown in RCA Victor service manual. *Courtesy David Sarnoff Library.*

Two-stage Audio with Push-Pull Output

The higher-priced models of 45 rpm phonograph contained two-stage amplifiers with push-pull output, driving either an eight-inch speaker or a six-inch woofer and four-inch tweeter. Typical output tubes are 35L6s or 35C5s. These are the best sounding amplifiers that RCA Victor used on the 45 rpm machines.

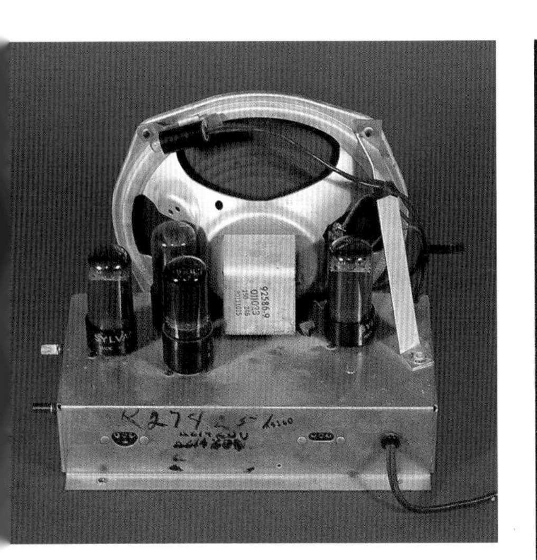

Model 45EY4 amplifier. Models 45HY4 and 8EY4 are similar except for tube complement.

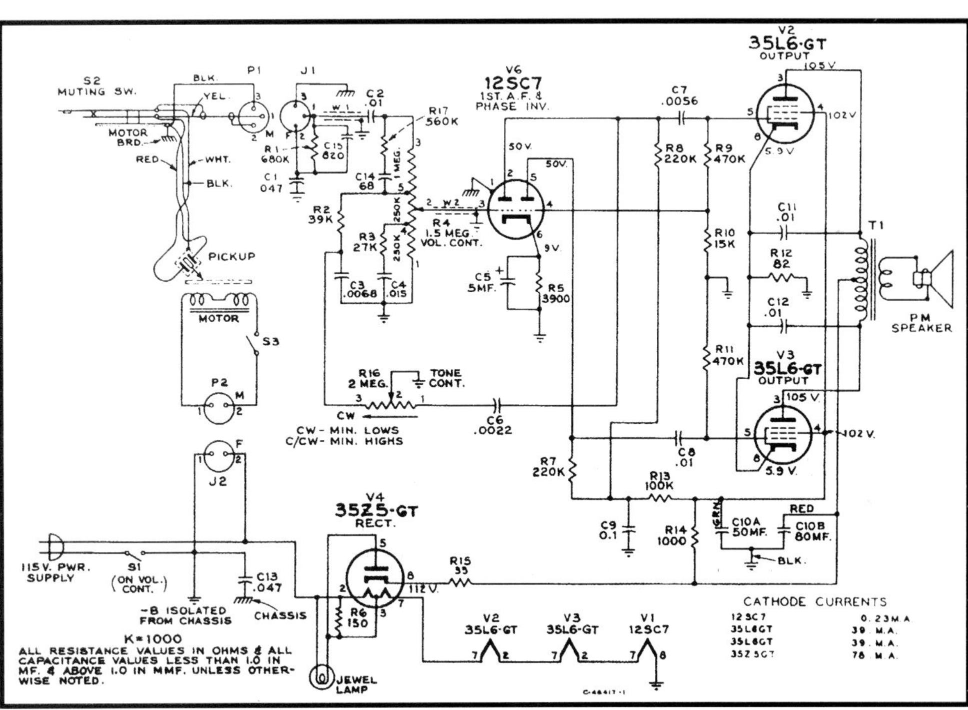

Schematic for push-pull audio amplifier model 45EY4 as shown in RCA Victor service manual. *Courtesy David Sarnoff Library.*

Model 7HF45 and 8HF45P amplifier.

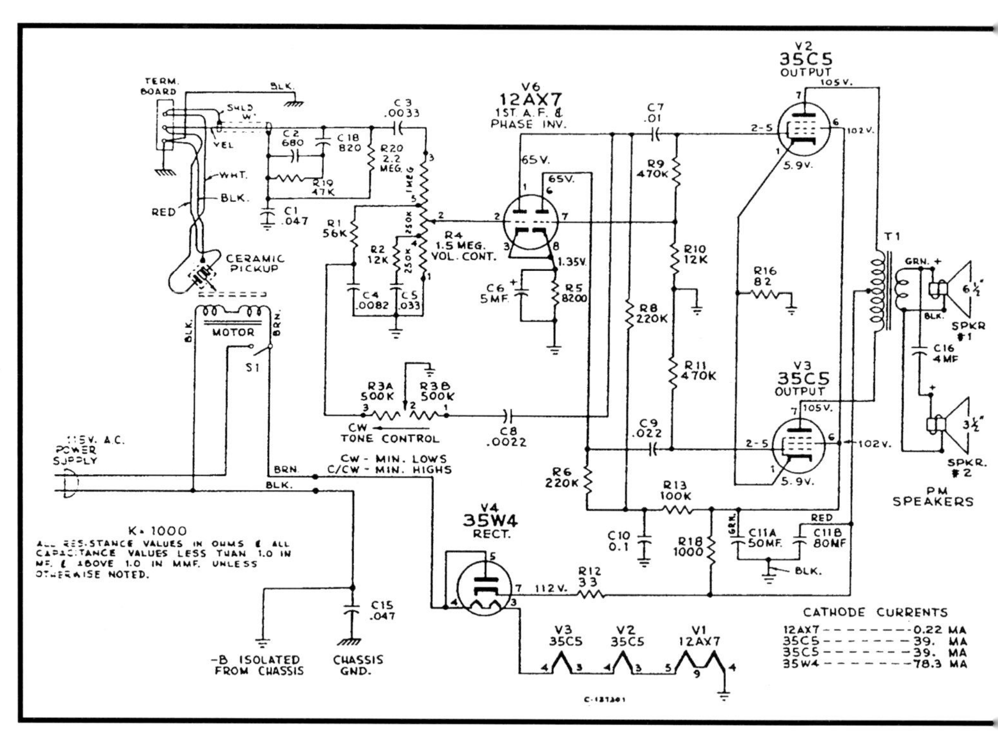

Schematic for push-pull audio amplifier model 7HF45 as shown in RCA Victor service manual. *Courtesy David Sarnoff Library.*

Radio/Phonograph Amplifiers

Table model radio/phonograph amplifiers are AM only. Three different tube layouts are used. Some five-tube super heterodyne receivers are employed with miniature tubes. The tube complement is: 35W4 (rectifier), 12BE6 (converter), 12BA6 (IF amp), 12AV6 (detector, AVC), and 50C5 or 50L6 (output). Other sets used full size tubes with the following tube layout: 12SA7 (converter), 12SK7 (IF amp), 6AQ6 (AF), 6AQ6 (Detector, AVC), and a pair of 35L6 (output). Sets designed to work from batteries as well as AC power have the following tube complement: 1R5 (converter), 1U4 (IF amp), 1U5 (detector, AVC), and 3V4 (output). Full size consoles had much more sophisticated circuits and many included FM reception and television. These chassis and tube complements will not be included in this book.

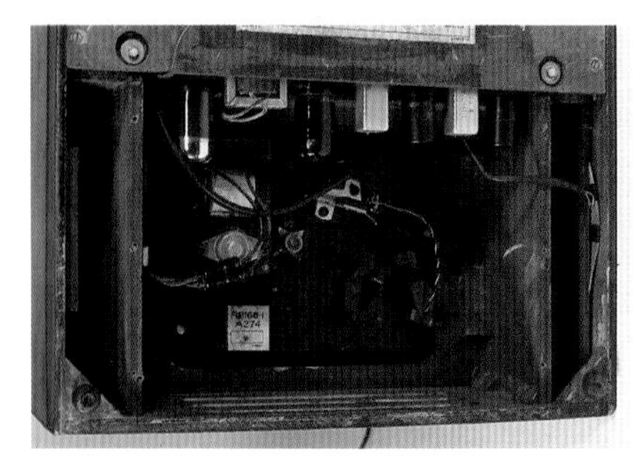

Model 9Y7 receiver amplifier.

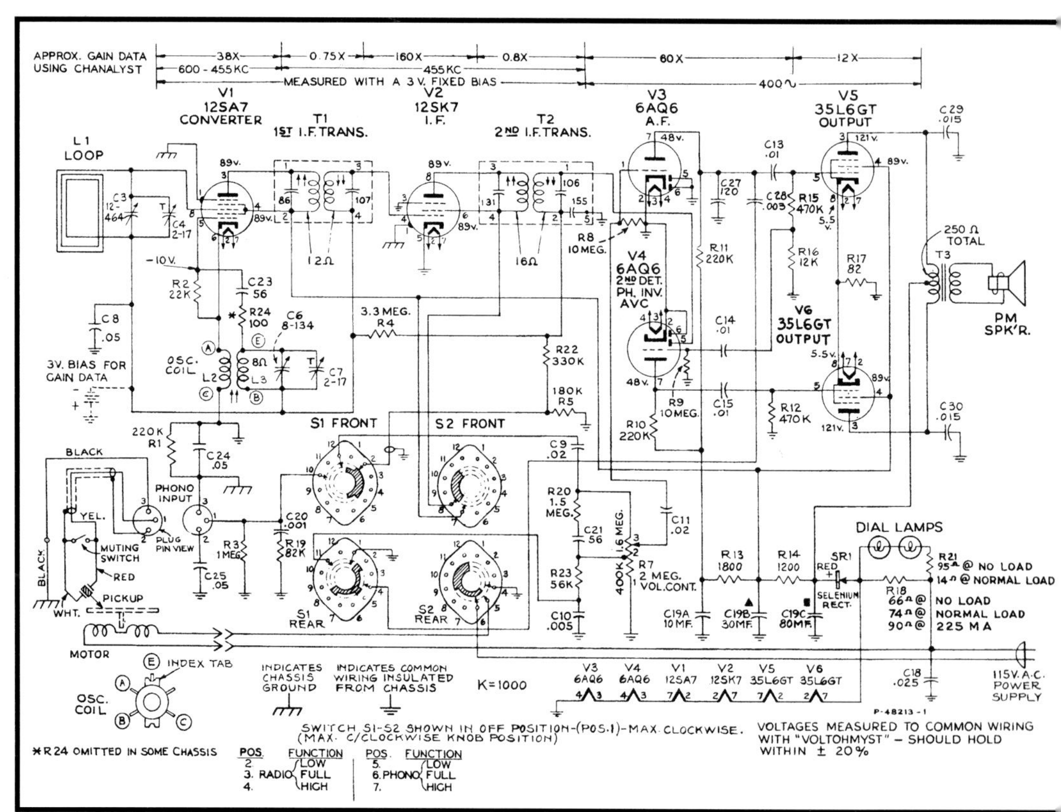

Schematic for radio-phonograph model 9Y7 as shown in RCA Victor service manual. *Courtesy David Sarnoff Library.*

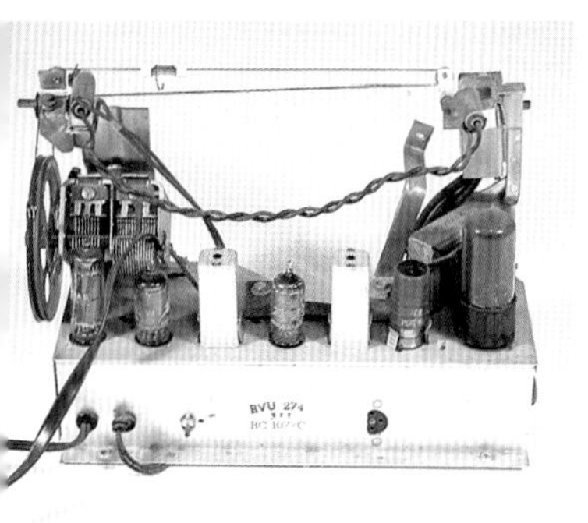

Model 9Y510 receiver amplifier.

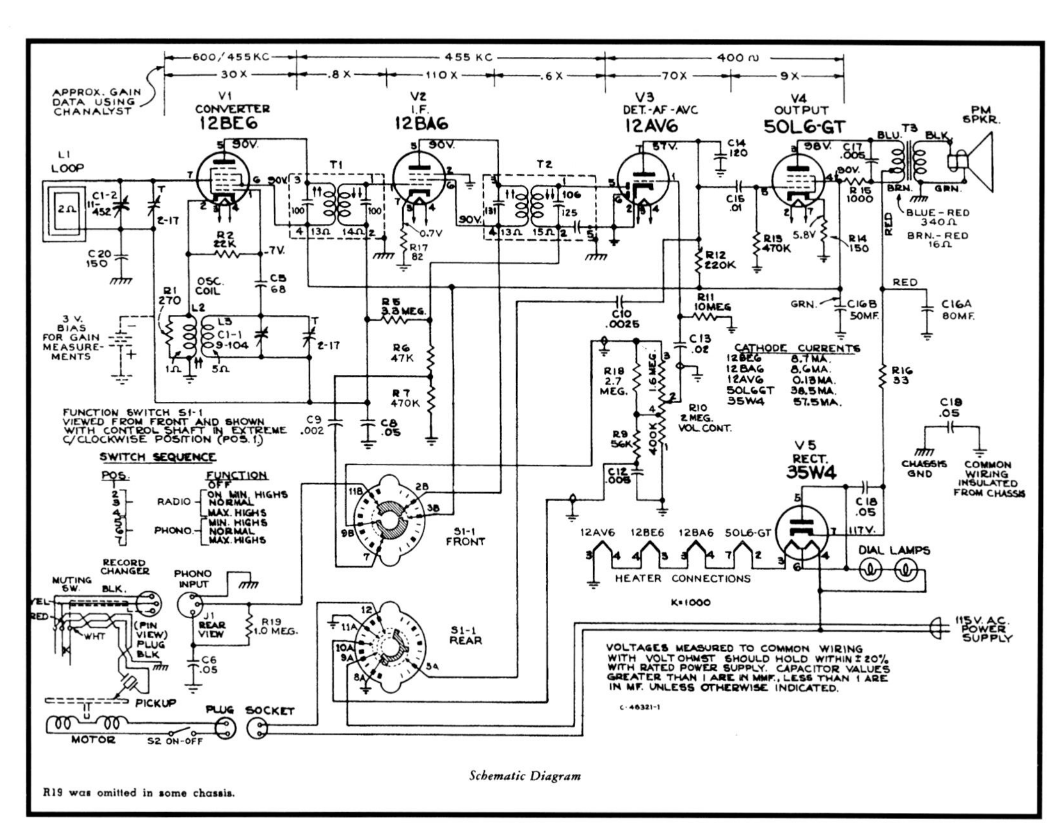

Schematic for radio-phonograph model 9Y510 as shown in RCA Victor service manual. *Courtesy David Sarnoff Library.*

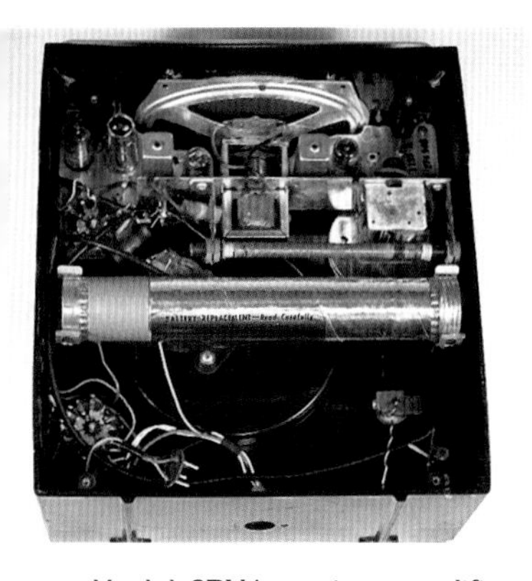

Model 6BY4 receiver amplifier.

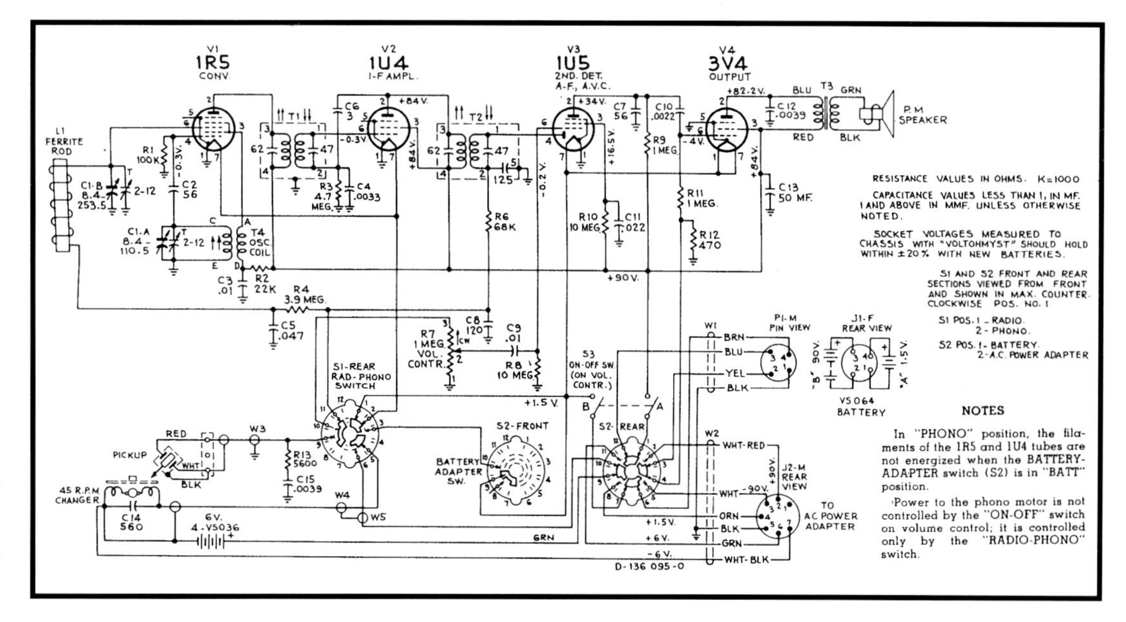

Schematic for radio-phonograph model 6BY4 as shown in RCA Victor service manual. *Courtesy David Sarnoff Library.*

Rebuilding Amplifiers and Radio Receivers

Warnings!

•Mold and mildew can grow on the chassis. It usually resembles a white or green powder. If you want to try to remove it, be careful not to stir it up in a dust cloud and breathe it. A severe allergic reaction can result.

•Selenium rectifiers were prevalent in the 1950s, and can be recognized by the way they look like accordion plates. When these rectifiers fail, they can emit toxic fumes.

•You should have a basic understanding of electricity before you attempt to repair these devices. Under certain conditions, the voltages inside these amplifiers could injure or even kill you. Make sure you follow a few simple rules for safety's sake.

1. Whenever possible, remove power before working on the amplifier.
2. Electrolytic capacitors will store charge for a considerable time even after power has been removed, so avoid probing near these components after removing the plug from the A.C. outlet.
3. If you need to have power applied and are probing the set, wear sneakers, don't stand in water, keep one hand in your pocket and do not lean on any metal surfaces. This way if you come into contact with any voltage, it will affect only your fingers, and will not travel through your body.
4. If you are working in a damp or wet basement, I suggest you take up another hobby!

Basic Cleaning Procedure

Remove tubes one at a time, so you don't get confused about where they plug in. Older tubes with Bakelite bases should be pulled out by the base, and not by the glass bulb. If the glue used between the base and the glass fails, you will be pulling the glass bulb out of the base, and the tube will be useless. Some sets use a metal clamp to hold the tube in place, especially when the tubes are hanging upside down. Push the metal clamp towards the chassis. This will disengage the clamp from the base of the tube. While you are holding the clamp in this position, use your other hand to gently pull the tube out of the socket. Resist the temptation to clean the glass bulb of the tube with glass cleaner. Many times it will remove the number markings from the bulb. Use a slightly damp cloth, and clean the glass parts that have no numbers. Then carefully wipe the numbers with a dry cloth.

Use compressed air to blow out any dust from the chassis. If a radio is present, blow out the variable capacitor. This is the device that has the tuning knob connected at one end and the other end has metal plates that slide into each other as you turn the tuning dial.

Rebuilding

With electronic devices that are fifty or more years old, the most common problems are leaky capacitors and dirty controls. Tubes and resistors are less likely to present a problem. In order to do a thorough rebuilding job, it is imperative that all waxed paper caps are replaced, along with all electrolytic filters. While this might seem like a daunting job, there are some tricks that can save time. Many of the new capacitors are no longer polarized, so you don't have to worry about the configuration. Much time can be saved simply by cutting the old capacitor out, leaving the leads intact. Now connect the new capacitor to the old leads and solder. Do this procedure one capacitor at a time to minimize mistakes. Another tip is to curl the new capacitor lead and wrap it around the old lead wire. A wire wrap tool or gun does this very quickly and neatly, but it can be done also by wrapping the wire around an appropriate size drill bit.

Modern replacement capacitors are more compact. The first two are electrolytic filters for the power supply. The last capacitor on the right is a typical coupling capacitor that is used to couple two amplifier stages together.

New replacement capacitor shown installed on old capacitor's leads.

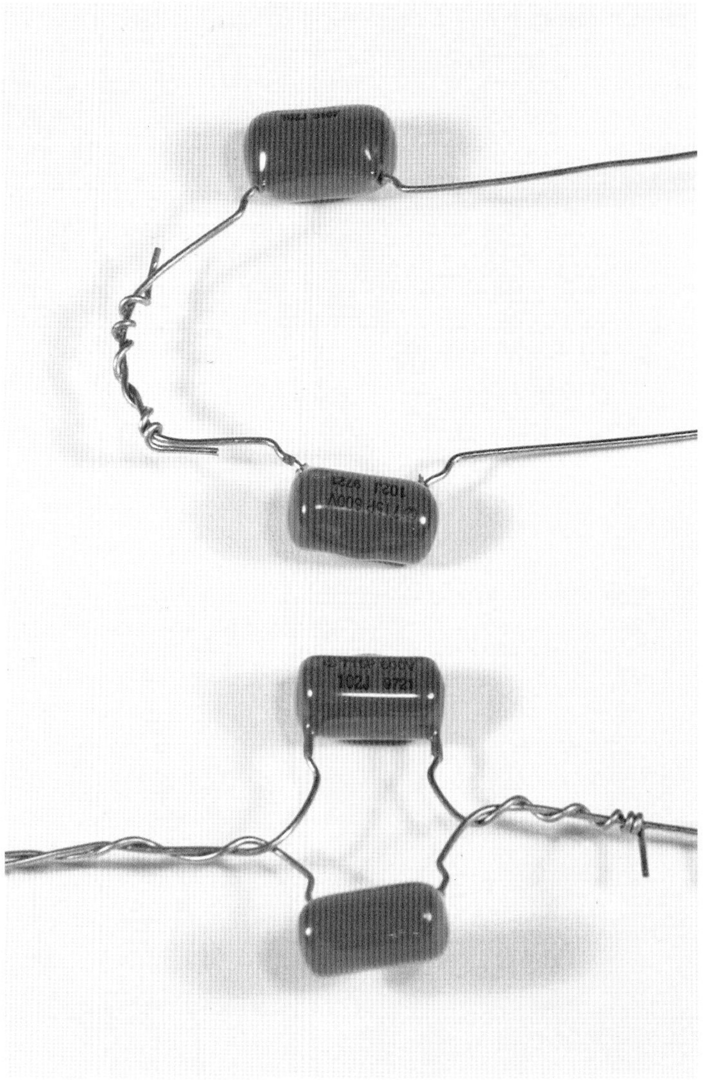

Changing capacitor values by putting them in series or parallel.

The filter capacitors are usually grouped in a common can made of metal or cardboard. The new, smaller, single-filter caps can be soldered in wherever it is convenient on the chassis. The old filter wires can be left, after cutting out the old filter can. If the can is mounted on the chassis, you are better off leaving the can in place and wiring the new filters under the chassis. Make sure that you disconnect the old filters when hooking up the new ones. Otherwise, problems can occur down the road if any of the old filters short out. Filter caps are polarized, so make sure to connect them correctly. If you can't find the exact value for a new capacitor to match the old one, choose a capacitance value that is close to the old one. The voltage rating must be at least the same as the old rating, or it can be higher. Capacitor values can also be joined together if you remember the following rules: Capacitance values add when they are connected in parallel. In other words, .01 mfd capacitor in parallel with another .01 mfd capacitor equals a value of .02 mfd. If you put the same capacitors in series, the total value would be half of .01 mfd, or .005 mfd.

Squirt some electrical contact cleaner into all switches and controls, and work the controls back and forth to distribute the cleaner. Next, check to see if any of the resistors are discolored or burnt looking. If you find one like that, you should replace it. Gently remove all vacuum tubes and clean the contacts with contact cleaner. Make sure that the tubes are cool to the touch before you do this. If a radio receiver is involved, clean the variable capacitor. Put a dab of light machine oil on the pivot points, and blow out all dust and dirt from the fins. Be careful not to bend the fins. Also lubricate any rollers that carry the dial cord. Make sure you don't get oil on the dial cord. If there is a radio bezel that displays the various radio frequencies, remove it and clean both sides of it. Care must be taken on the side that contains the lettering. In order to avoid accidentally removing some of the lettering, you might want to wipe that side with a dry cloth only.

Double check to make sure that you have soldered all connections under the chassis after your recapping job. At this point you will want to power up the chassis and see if it works properly. First, however, I suggest making a simple tool that will protect you from burning out parts, in case you made a mistake. Take a lamp socket and connect it in series with a short extension cord as shown. Install a 100-watt lamp. Now plug in the chassis. If there are any shorts in the chassis the lamp will light brightly. If the lamp barely lights, or doesn't light at all, it means the chassis is not drawing excess current. Most chassis will play with the lamp in series.

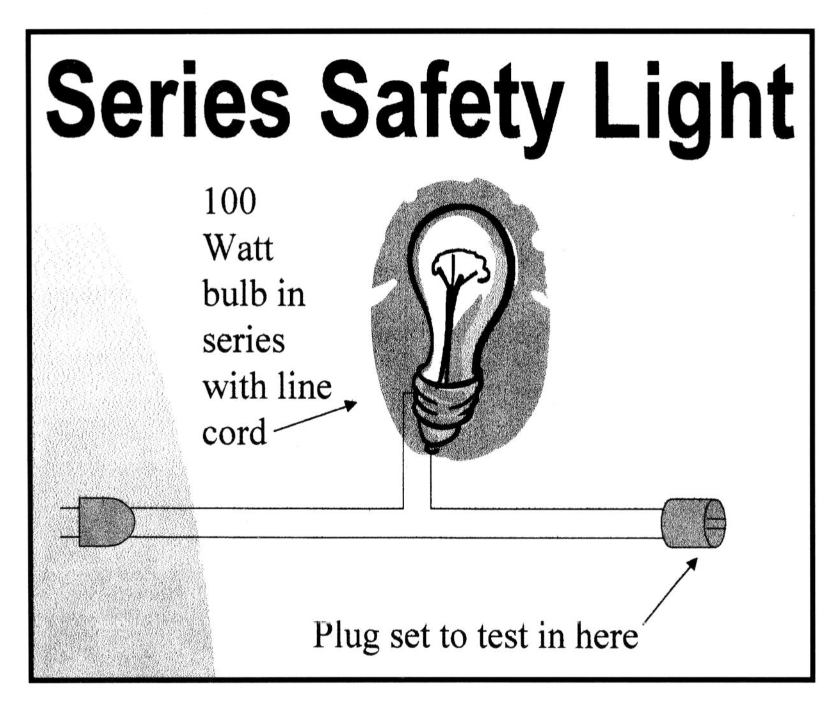

Anatomy of a safety test cord.

Although about 90% of amplifiers and radios will spring back to life after this work is done, there are some units that will still have problems. The best thing to do if the set is still not functioning is replace the vacuum tubes one at a time. If you don't have spares available, have the tubes tested. This used to be a simple task of bringing the tubes to your local pharmacy but today you will have to find a repair shop that is equipped to fix older sets. If the tubes are not the problem, then a knowledgeable technician who is familiar with schematics and test equipment should troubleshoot the set.

Cabinet Restoration

The following information comes from years of trial and error restoring many cabinets. Many of the models shown in this book were restored using these techniques. Keep in mind that there are many different ways to attack these problems and these are just my preferred methods. There are probably many other excellent methods also.

Bakelite Cabinets

Clean with soap and water and a soft bristle brush for the nooks and crannies. After drying, apply Gojo cream hand cleaner and work it into the surface. Wipe it off with a clean cloth. If the Bakelite outer shell appears to be in good condition, apply Novus #2 cleaner and polish. Start in an inconspicuous spot on the back or side and see how the Bakelite reacts. If it yields a mirror-like shine, proceed. If you see a light brown condition where you applied the Novus, STOP. The Bakelite shell is in questionable condition, and abrasives should not be applied. Apply a coat of good quality pure carnauba wax. If the Bakelite case is in very poor condition, an application of dark brown shoe polish can improve its appearance.

Wooden Cabinets

Whenever possible it is always more desirable to work with the existing finish rather than refinish. It is very difficult to make the refinished wood look like the original finish. The following procedure typically can be done effectively only once, because some of the original finish will be removed. As long as the original finish is intact, the dull outer shell can be removed yielding the original sheen.

First use "0000" steel wool and work with the grain of the wood. Don't overdo it. The objective here is not to remove the finish, but only to smooth it out and polish it. Clean off the dust produced by the steel wool. Fill in any nasty scratches with a scratch cover pencil that they sell in hardware stores. I prefer the type that has a felt tip and not the grease pencil type. Apply Novus #2 and buff the finish with a clean soft cloth (old undershirts are my favorite). Finish off with a coat of pure carnauba wax. This can be done on all colors of finishes—dark or light.

Vinyl Covered Cabinets

These cabinets are the biggest challenge. If the vinyl is scraped up it really can't be repaired fully. There are some tricks that improve the appearance, however. Glue down any frayed vinyl, especially on the corners. Touching up the vinyl with colored shoe polish is also a useful technique. Every color you can imagine is now available at shoe repair stores you'll find in shopping malls. Cleaning the forty- to fifty-year-old vinyl is also a challenge. If a good grade of vinyl was not used in the phonograph's cover, the cleaner may damage it. Start by applying a moderate strength cleaner like Formula 409 on a clean cloth and apply it to the vinyl. Wipe off quickly and check the results. If no discoloration is observed, and the dirt is being removed, you are making progress. Be extra careful with vinyl patterns and darker colors. The objective is to remove the dirt but not the color or patterns. Some vinyl is so thin that the cleaner seems to soak through when you try to clean it, and looks worse when you are finished. This is especially prevalent on the vinyl used inside the lid of many portables.

Once the vinyl is clean, you can treat it with a vinyl protectant like Armoral.

Another problem prevalent with the vinyl-covered portable players is mildew. As mildew starts to grow, it gets inside the wood behind the vinyl, making it almost impossible to clean. There is a solution, however. Place the cabinet in a large plastic garbage bag with a plate containing plain kitty litter (no deodorizers). Close up the bag tightly and leave it for a week. When the bag is opened, both the smell and the mildew will be gone. I have done this to several units, and the smell has not returned.

Chapter Fifteen

Manufacturer Dating and Service Tips

Dating RCA 45 Players

by Bob Havalack (Eager Brothers Music)

Thus far, this book has indicated when the RCA Victor 45 players were designed, announced, and introduced for sale. To determine when a particular player was manufactured, one can refer to a date code that RCA Victor used for the players. RCA, as did many manufacturers, used a date coding system to track warranties and production totals.

Bottom view of rp-168 motorboard showing date code 915 (15th week of 1949).

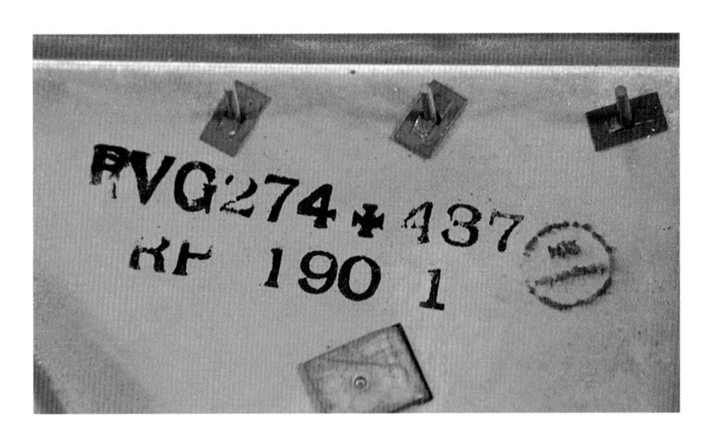

Bottom view of rp-190 motorboard showing date code 437 (37th week of 1954).

RCA chose a year/week coding system. The coding consists of three characters. For example, in the date code 852 the number "8" was for the year 1948 and "52" was for the 52nd or last week of 1948. This is when the first 45 rpm players were manufactured for demonstration purposes. Product reviews were then used to launch the players. Date codes were applied to many RCA products, including Victrolas, radios, televisions, parts, and so on. RCA was assigned the number "274" by the Electronic Industries Association (EIA) to identify RCA products.

I have chosen at random a few 45 rpm players to use as examples in locating the date codes.

9EY3

Turning the player over and looking into its bottom, either a white paper tag or a white ink stamp can be seen, located on the motorboard. The particular player used here as an example has the white tag and it reads RP-168 A274 933. The RCA date code system indicates that the motorboard was produced during the 33rd week of 1949. By removing the bottom cover and observing the electrolytic capacitor (tall cardboard tube mounted on chassis) one can see the following numbers: 235925, in which the 235 is the EIA code assigned to Mallory, the manufacturer of the capacitor, and the 925 is the year and week of manufacture, or 25th week of 1949. Moving on to the volume control, the numbers 137920 can be seen. Once again, the last three digits date the part: the volume control was made during the 20th week of 1949. Pulling the three RCA tubes out reveals that each tube has the same single number, hyphen, and two additional numbers (9-31). These tubes were produced during the 31st week of 1949.

45EY3

Removing the bottom cover, a group of numbers can be seen in the upper left-hand corner of the motorboard. They read RP-190 K274234. 274 is the EIA number assigned to RCA and 234 tells us that the changer was made during the 34th week of 1952. Moving on: the output transformer has tm1 230 stamped on it. This transformer was manufactured in 1952 during the 30th week. When the amplifier is removed from the cabinet the numbers 274232 can be seen. It was manufactured in the 32nd week of 1952. The volume pot has a stamp on the bottom with the numbers 104227. Manufacturer code is 104 and the production date is the 27th week of 1952. The electrolytic capacitor is marked 472226. The 472 is the EIA number assigned to Pyramid (manufacturer of the capacitor) and date of manufacture is the 26th week of 1952. Other capacitors are checked as well, and the finding is that all were produced during the 26th week of 1952. Now the original pickup cartridge is removed, to reveal the following numbers: 274224 and 74067. RCA is the manufacturer and the date is 1952 during the 24th week. 74067 is the RCA part number assigned to this cartridge. (RCA used a number of different cartridges in the players).

6BY4

The first thing noticed on this player upon removing the bottom cover is that an actual date is stamped in red ink under the battery tube. It reads "Sept 8, 1955." Actual dates are found on most 6BY4 players, and here and there on players with production dates starting in the mid 1950s. The tubes in this player are all dated 5-30, indicating the 30th week of 1955. The chassis and the model tag both read 540, indicating the 40th week of 1955.

45J2

This player was found in its original shipping box. The shipping box is dated 10-24-50 (October 24, 1950). The cartridge is stamped 043 indicating 1950, 43rd week. The changer is stamped 043, also indicating the 43rd week of 1950. Using a 1950 calendar and counting the full weeks, one can see that the 24th of October indeed occurs during the 43rd week of 1950.

The restorer should always keep in mind that a player may have been altered before he or she received it. The changer may have been exchanged with that of another unit, the cartridge may be a newer Astatic replacement, and all of the capacitors may have been replaced at some time.

Service Tips

Since 1995 the author has been writing a quarterly newsletter dedicated to the 45 rpm machines called *The 45 RPM Phono Gazette*. One of the features of the publication is a column called "The Workshop," wherein I discuss techniques developed through years of experience working on these machines. Below are some of these tips.

Tip: Motor Bushings

OK, you just picked up a nice 45 rpm unit. What's the first thing that you notice when you take the bottom cover off? The motor is about to fall into your lap because the rubber bushings have disintegrated. When the motor is in this state, the idler wheel will not engage the turntable properly. Secondly, the motor itself can rub against the motorboard, causing mechanical noises to enter the amplifier. I have frequently scavenged motor mounts from a record changer from the 1970s or 1980s or have used grommets that were not really intended for this application, but are sized similarly to the originals. Today, however, it is possible to buy the correct bushings from several sources. Refer to Appendix C for resource information.

Tip: Noisy Amplifier Fix

One of the deficiencies of the amplifier used in the model 45EY4 is the use of a 12SC7 tube as a voltage amplifier. Apparently these tubes are quite noisy. Luckily, there is a simple way to improve the set's performance. Doug Houston suggests that a 12SL7 tube can be substituted, if some simple socket connection changes are made. Any old RCA Tube Manual will provide the restorer with the pin conversions.

Tip: Removing Gook From Plastic Tonearm

Here's another tip for cleaning all that stuff that you usually find on the plastic tonearm after the previous owner weighed it down with pennies, nickels, and/or quarters. The remnants of forty-year-old tape can be cleaned off by soaking the tonearm overnight in soapy water (start off with hot soapy water). The next day the gook should easily come off with your fingers. You can get into the crevices with a soft brush. This is the only safe way to clean the soft plastic arm used with the rp-190 changer. Finish it off with a good polishing using Novus #2 Plastic Polish.

Tip: Fixing Rejectitis

Rejectitis! That's my term for the condition characterized by an Rp-190 record changer's continuously rejecting. It usually happens after some records have been played, and suddenly the changer seems to have a mind of its own. Rejectitis is caused by the rubber cycling cam coming too close to the rotating spindle that contains the knurled roller gear. When the reject cycle is complete, the notch on the cycling cam should be facing the knurled roller gear. If they are too close, the momentum of the rotating cam does not allow it to stop in time, and the cam re-engages after the notch has passed the knurled roller gear. Luckily there is an adjustment for this. Loosen the bracket that supports the slider underneath, and move the bracket toward the center spindle. Then tighten the bracket back down. If you have moved the bracket too far, a different problem will occur: the tonearm will not reject at the end of the record, because the cycling cam did not finish its cycle. In other words, the notch is not facing the knurled roller gear. A happy medium must be found between these two conditions. It might take a couple of adjustments to get good results, but once the optimum adjustment is found, "rejectitis" will not be a chronic problem in your phonograph's operation.

Another cause of "rejectitis" is produced when the record changer is not on a level surface. If the left side of the changer is lower than the right side, this condition can cause continuously repetitive cycling. Make sure that the changer is level before making the slider adjustment.

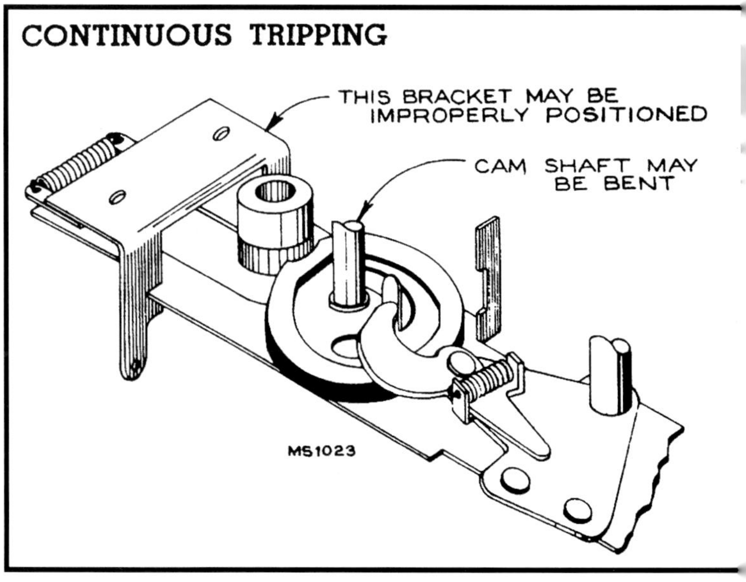

CONTINUOUS TRIPPING

THIS BRACKET MAY BE IMPROPERLY POSITIONED

CAM SHAFT MAY BE BENT

MS1023

Rp-190 slider showing interaction of parts. *Courtesy David Sarnoff Library.*

Tip: New Old Stock Parts, International Models

Over the last few years I have managed to acquire some new old stock (NOS) parts. Included are idler wheels, cam wheels, tonearms, a motor, assorted spindle caps, shelves and separators, grommets, etc. Pictured with the NOS stuff is a 6EY1 International model. For those readers who may not know, international models were not permitted to display "Nipper." In its place is either the word "Victrola" or the acronym "RCA." On the 6EY1 the "Nipper" emblem on the tonearm is replaced by "RCA." In my collection are three international models: the 6EY1 shown here, a 45EY3, and an 8EY4FQ.

These models also have a voltage switch under the turntable that permits 115 or 230 volt operation.

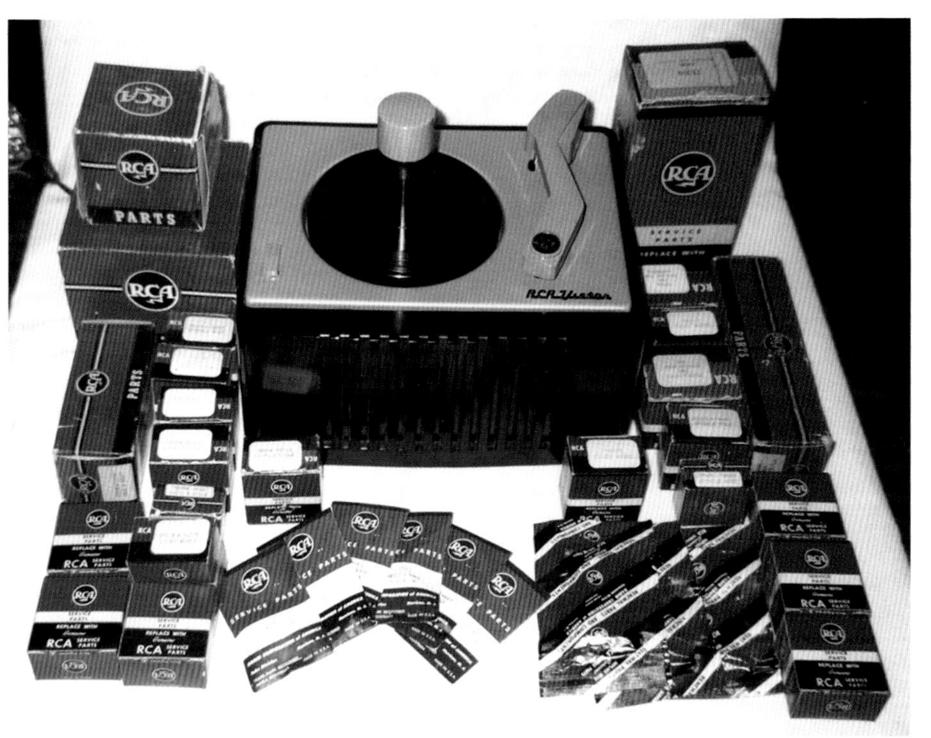

Model 6EY1 surrounded by collection of NOS 45 phonograph parts from RCA Victor.

Collection of NOS 45 phonograph parts being watched over by "Nipper."

Tip: Speaker Problems

When servicing these machines, sometimes the speaker presents difficulties.

If the sound seems fine at moderate to loud volume, but seems distorted at lower volume, the speaker voice coil alignment is usually the problem. At the lower volumes, the voice coil does not move freely. Although the speaker could perhaps be fixed by a professional technician, my recommendation is to replace it. Speakers are still readily available, but the oval types used on some models are more difficult to replace because the new speakers have larger magnets than on the old speakers. I actually did fit a Radio Shack oval speaker into a model 45EY2 after repositioning some of the components in the amplifier. The 45EY3 is another story. The only solution, if a NOS replacement is not available, is to use a four-inch circular speaker. Not very elegant, but the new four-incher doesn't sound much different than the original oval speaker.

Another symptom of a bad speaker is vibration or distortion when the sound level is increased. First check to see that it isn't something else vibrating loose. I usually remove the amplifier and the speaker from the cabinet and test again for vibration noise. If the vibration is still there, the speaker is the culprit.

The strangest speaker problem I have ever seen was one in which it was impossible to get any bass output from a 7HF45. The speaker was removed, revealing cardboard that had been wedged between the speaker and the baffle! With the cardboard removed, the bass was back!

Tip: Static and Noise in Amplifiers

A fairly common problem encountered when testing an old tube amplifier is noise and static. Even after the amplifier has been recapped it can still persist. Many times the noise is caused by poor connections at the tube pins and socket. While the noise is heard, gently rock each tube from side to side. If the noise becomes worse during the rocking, the connections are faulty. Remove each tube and, after they have cooled down, spray the pins with contact cleaner. Then insert the tube. Remove and re-insert it several times. This process will clean the pins and sockets. Now the amplifier should be nice and quiet.

Tip: RP-190

As a record changer ages, the shaft on which the tonearm swivels can get sloppy or loose. This can lead to the following condition: When the machine starts to reject at the end of the first few records on the stack, the arm does not lift to the required height. As the arm swings away from the record, its height increases, and it finally reaches the maximum height when the arm is ready to descend on the next record. The arm should lift to the required height and stay there during the rest of the reject cycle until it descends to the next record. To fix this condition, add another washer to the shaft under the tonearm. This will take out some of the extra slack that has developed there.

Upper right corner shows tonearm mounting of rp-190. Adding another washer under the "C" clip can help.

Tip: Quick and Cheap Record Changer Stand

Here is a helpful tip from Bob Havalack in Rochester, New York. When working on the rp-190 changer, screw in a few three-inch bolts that have the same threads as the bolts that were extracted to remove the turntable from its cabinet. With the three-inch bolts attached, the restorer can quickly run the changer outside of the cabinet and flip it over for adjustments.

Tip: Arm Not Coming Back To Start Next Record

If the tonearm on the rp-190 does not return to the lead-in groove of the next record, the return spring linkage requires cleaning. Paul Childress suggests disassembling the parts and cleaning the shaft with fine emery paper until it shines like a penny. Another area that Paul feels must be super clean is the return spring arrangement for the idler wheel. He removes the plate assembly on top of the motor and thoroughly cleans it with Gojo cream hand cleaner and water.

Spring in upper left corner is what forces tonearm to return to the starting groove of the record.

Return spring and other parts have been removed. Clean the remaining shaft thoroughly.

Rp-190 changer standing with three-inch screws supporting it.

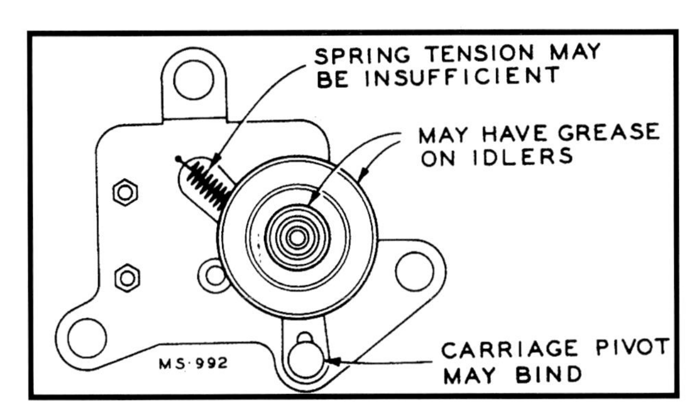

Clean and lubricate the carriage pivot in the diagram. Use light machine oil or very light grease. *Courtesy David Sarnoff Library.*

Tip: RCA Victor Tubes Not All the Same

Ever notice that some push/pull amplifiers sound better than others? Bill Jones has found that earlier RCA output vacuum tubes with gray lettering sound better than the newer tubes with red lettering.

Tip: Base Vibration Fix

Bill Jones has also found a way to stop vibration problems on model 8HF45P. Bass output causes the grill cloth to vibrate against the wooden support slats where the woofer is mounted. Remove the woofer and slide half of a Popsicle stick with some glue between the grill and the wooden slat. Deposit a little bit of the glue so that the grill will stick to the wooden slat. Remove the stick and reinstall the woofer.

Tip: The Storage Problem

Collecting things that are of particular interest to the hobbyist can be quite rewarding. There is one aspect of the hobby, however, that can cause big-time headaches: the problem of storage. Most folks who start collecting have no idea what they are in for once they get into full swing. Whether collecting automobiles, phonographs, or thimbles, the need for space will eventually become an issue. "If only I had a basement or a bigger house" is often heard. The sad thing is, if the collector were to dig out a basement or move to a bigger house, he or she would fill it up in short order and be in the same predicament. Now that I have a sizable collection of phonographs and radios, I have made it a quest to make ingenious use of space and displays. If your place has eight-foot ceilings, you can make better use of the last foot or two near the ceiling by installing shelves in the garage or utility area. I make shelves from 24-inch wide doors, because people who live in my neighborhood are always discarding them. I've installed three of them above my workbench and one in my garage between the two overhead doors. As long as a person can walk under the shelves without hitting her or his head, that space has been maximized.

When it comes to displaying one's collectible gems, here are some things to consider. If the display shelf is wide enough, it can be set up like steps. In this way, items in back of each other will still be visible. Another eye-pleaser is displaying similar items (like 7EYs) at different heights.

Most of the units that I display are in restored condition, so there is another thing to consider when displaying them. The only way to keep a restored phonograph in good condition is to play it regularly. If you have only a handful of working phonos, playing them on a regularly rotating basis is not difficult. If, however, you have hundreds of phonographs, a system must be developed. Install AC extension strips (out of sight, if possible) and connect groups of six phonos to each strip. Every few weeks, turn on each group of phonos for about ten minutes. This procedure will prevent flat spots from appearing on the rubber idler wheels, and it also will keep the capacitors in good shape. Believe me, if a phonograph is restored and then allowed to remain idle for a year or two, there is a good chance that the idler will be ruined the next time the player is operated. Ironically, it turns out that most multispeed record changers do not have this problem because they are equipped with an "idle" position that takes the rubber idler away from the motor shaft when the player is not in use. Unfortunately, RCA Victor wanted to use the smallest number of parts possible in their new 45 rpm changer, so the idler always rides on the motor shaft, even when the player is in repose.

Part of the author's display room, showing use of shelves of different widths. All the units are plugged into AC strips so they can be periodically powered up—not all at once of course!

Tip: Rubber Cone Suspension Parts

As is the case with so many of the parts in these 45 machines, the rubber suspension cones are no longer available. Condition can vary from good to hard as a rock. Many are also found in a partially crushed state. A good rubber cone will be soft and pliable and will hold the record changer away from the base mounting so that any vibrations from the record changer will not be transmitted to the cabinet. If the rubber cone has hardened, it will transmit the vibrations to the cabinet and back into the amplifier, causing rumbling and other assorted unwanted noises. If the rubber cone has become partially crushed, the record changer will come into contact with the base, also causing the transmittal of extraneous sounds. The rubber cone directly under the motor is usually the first to crush, followed by the single rubber cone mounted under the tonearm. One way I have lengthened the life of many partially crushed cones is to mount a couple of washers between the record changer and the rubber cone. This way, the changer sits high enough to avoid contact with the cabinet base. As long as the rubber cone still has resilience, it will work. Many other record changer manufacturers used springs instead of rubber cones. Early rp-168 changers mounted in wooden cabinets used springs. Perhaps in the future we will find a good source for this type of spring and not have to worry about rotting rubber cones.

Chapter Sixteen

RCA Victor Memorabilia

For many many years, "RCA Victor" was quite a well respected name. The company displayed its name, and especially "Nipper," whenever it could. Some of these items refer to RCA Victor, while others refer specifically to the 45 rpm system of recorded music.

Vintage Nipper paper record bag. *Courtesy Rick Weingarten Collection.*

Nipper coffee mug.

Vinyl tote bag handed out at the 1956 Consumer Products Distributor meeting.

Nipper change pad or large coaster.

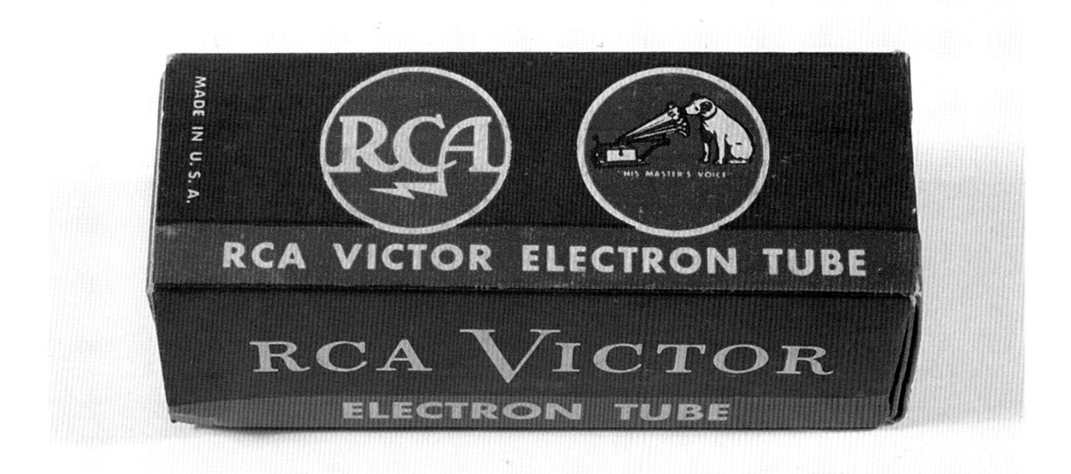

Vintage RCA Victor electron tube box.

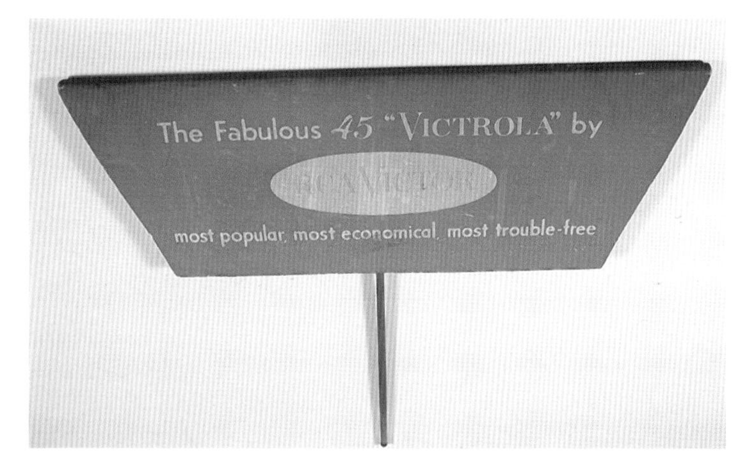

First side of vintage two sided display sign from RCA Victor dealer—"The Fabulous 45 Victrola."

Other side of vintage two sided display sign from RCA Victor dealer—"New Sensations in Sound."

Colored marbles featuring Nipper.

Stained glass Nipper Christmas ornament.

Wall lamp featuring Nipper.

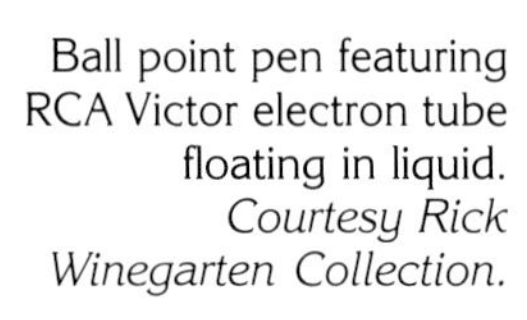

Ball point pen featuring RCA Victor electron tube floating in liquid. *Courtesy Rick Winegarten Collection.*

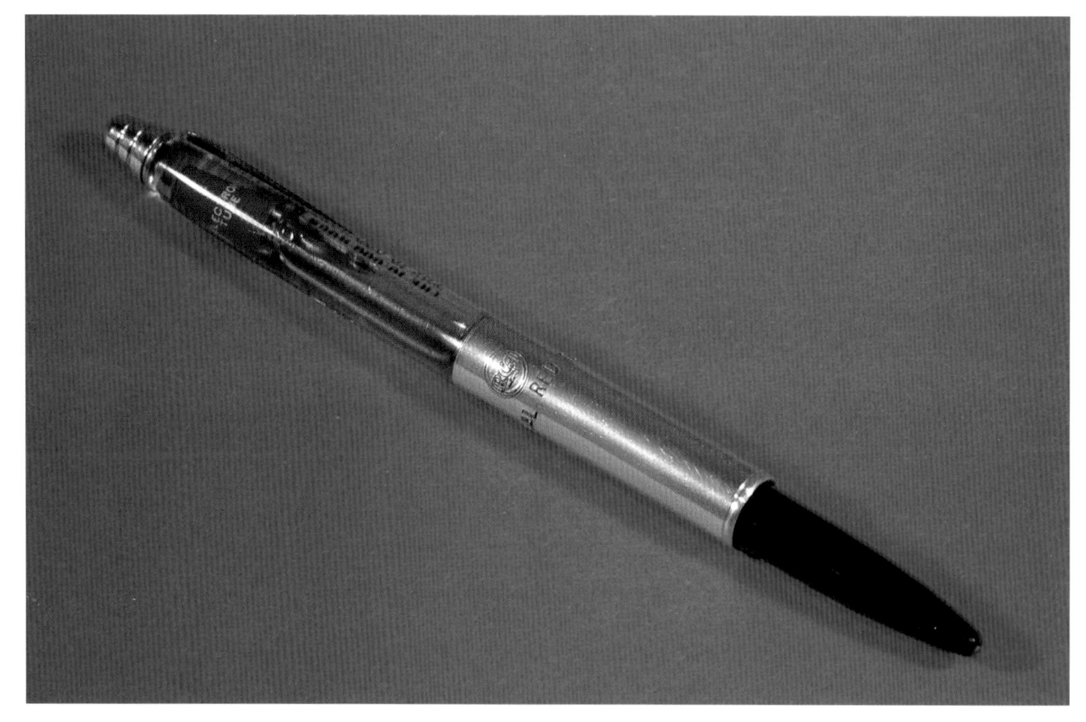

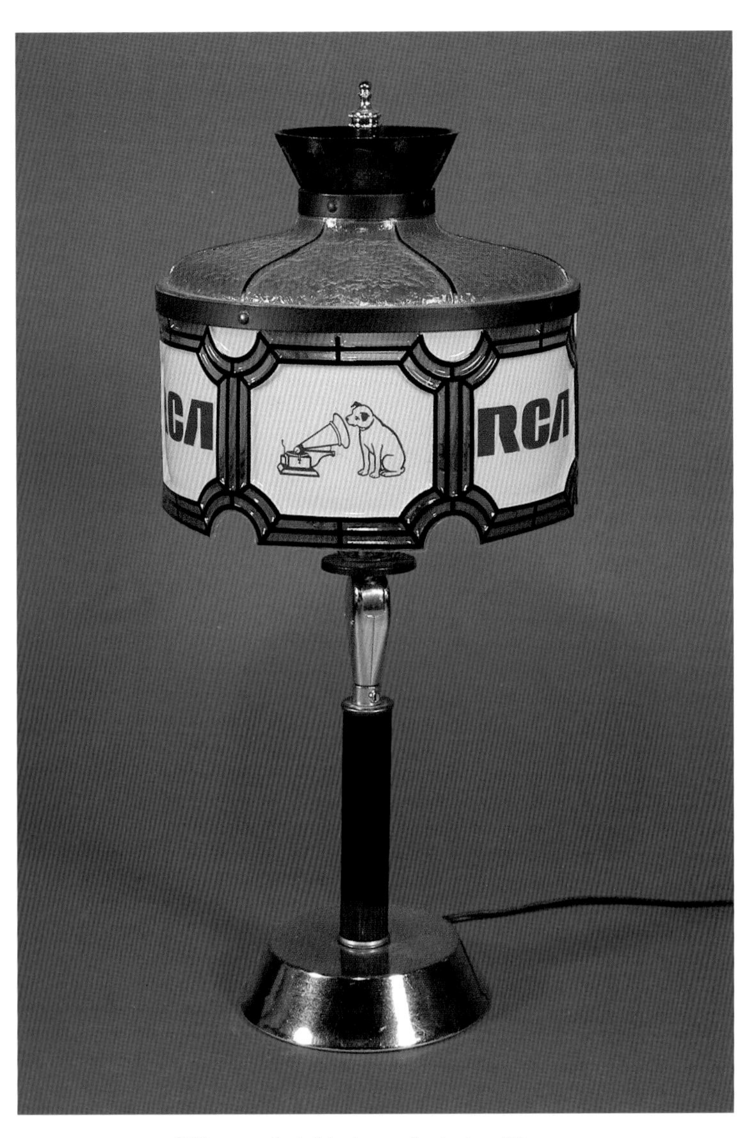

Tiffany style table lamp featuring Nipper.

Close-up of model 9EY3 on the blue boxer shorts!

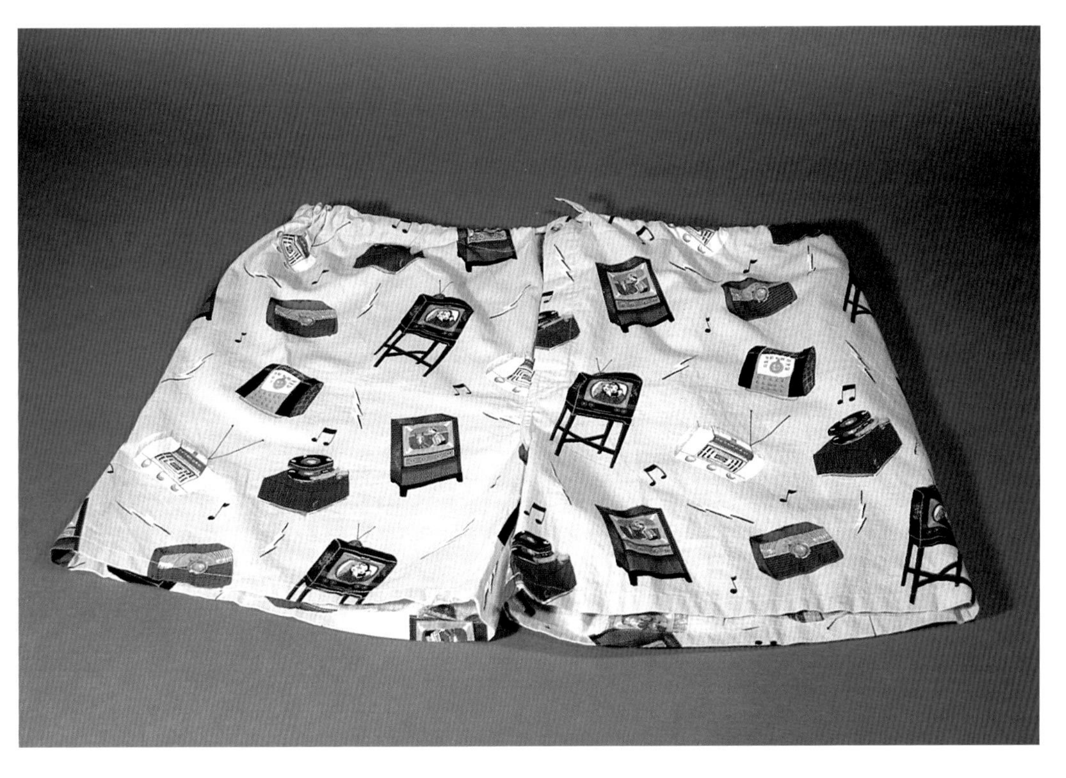

Blue boxer shorts featuring the RCA Victor 1949 model year. Includes model 9EY3. *Courtesy Ray Tyner Collection.*

1952 Calendar featuring color illustration of family enjoying music from console with 45 rpm system.

1953 Calendar featuring color illustration of family enjoying music on porch with 45 rpm system.

Front view of vintage match books. The first one advertises RCA Victor, the second one advertises the new "45" system.

Back view of vintage match books. The first one shows Nipper, the second one shows a color rendition of model 9EY3 playing red seal records.

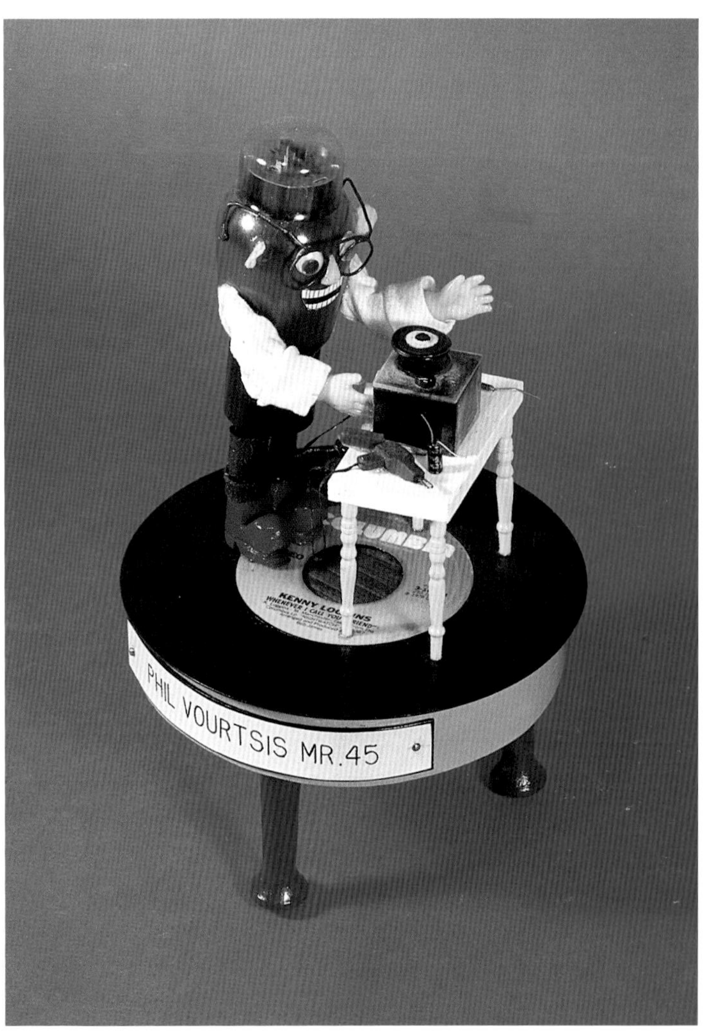

Custom made display piece showing the author repairing a 45 phonograph. *Custom work done by Bernie Gindoff.*

Assorted color key chain discs advertising the "45".

Adjustments

The following illustrations are all courtesy of the David Sarnoff Library.

RP-168 Series **ADJUSTMENTS**

Adjustment Sequence:

1. Synchronize separator shelf (Ill. No. 5) and separator knife (Ill. No. 5B) action (necessary only on rotating gear type of record separators).
2. Adjust position of star wheel (Ill. No. 62).
3. Adjust position of director lever (main lever) (Ill. No. 41) in relation to the star wheel by bending if necessary.
4. Adjust tone arm pivot screw (Ill. No. 12) for minimum side play without binding.
5. Adjust sapphire height above motorboard.
6. Adjust tripping position.
7. Adjust landing position.
8. Adjust pickup arm height during cycle.
9. Adjust position of muting switch so that contacts are open 1/32" during playing and are closed during cycle.

Separator Synchronization:

The following applies only to the rotating gear type of record separators:

1. Make certain the two embedded gears (5 and 6) are meshed with gear (7A) on the upper end of the star wheel shaft so the action of the separator knives is synchronized.

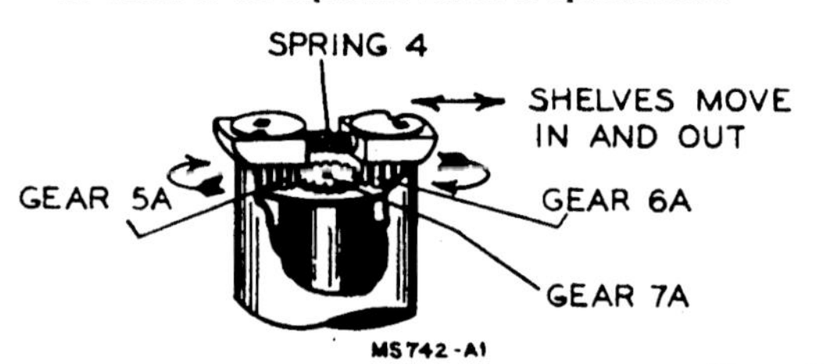

Figure 32.

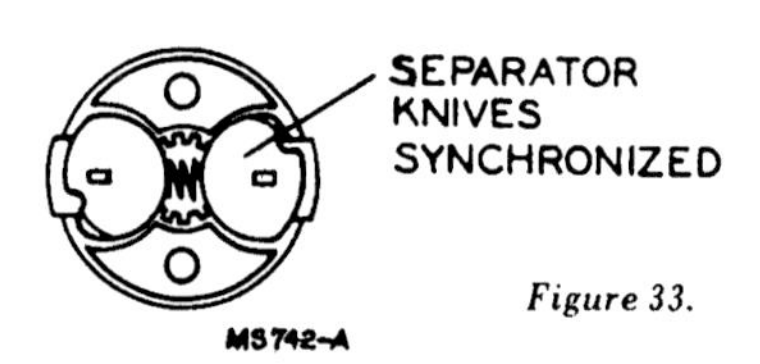

Figure 33.

Star Wheel Position:

1. Turn the star wheel so that the separator knives are in the position indicated in Figure 33 for rotating gear type of separators or fully retracted for push-out separators.
2. Loosen the two set screws (61) sufficiently to permit the star wheel to rotate without disturbing the shaft (7).
3. Rotate the star wheel points directly to a cam screw or nose screw (visible through slot) as shown in Figure 34.
4. Tighten the two set screws (61) and rotate the mechanism through a complete cycle to check operation. The separator knives must rotate 360° to the starting position as indicated in Figure 33.

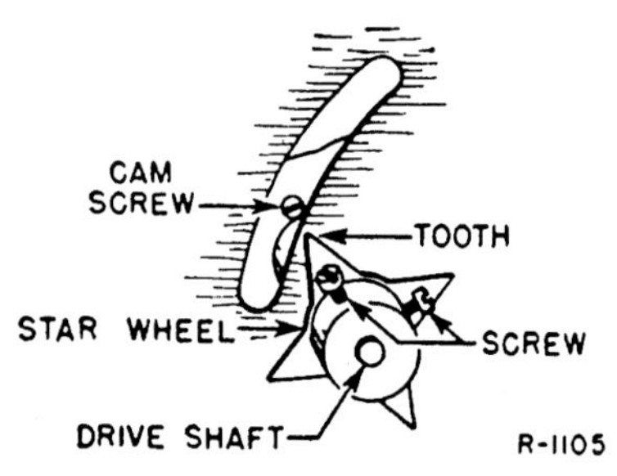

Figure 34—Star Wheel Timing.

Director Lever Position:

Push reject lever and rotate the turntable slowly by hand until the end (41C) of the director lever moves in to its limit of travel so when the star wheel is rotated it contacts by the amount indicated in Figure 35 for lever with long end. For lever with short end, the star wheel should first contact the end (41C) approximately 1 16-inch from the front or leading edge of the lever.

If the end of the director lever (main lever) is too close to the star wheel, it will jam. If too far away, it will cause erratic record dropping. If in doubt and unable to measure, move the end toward the star wheel until most of the play is removed when the star wheel is moved back and forth at this setting. With the push-out record separators and the lever with short end, there will be considerable play but the tension of the separator springs holds the star wheel against the lever.

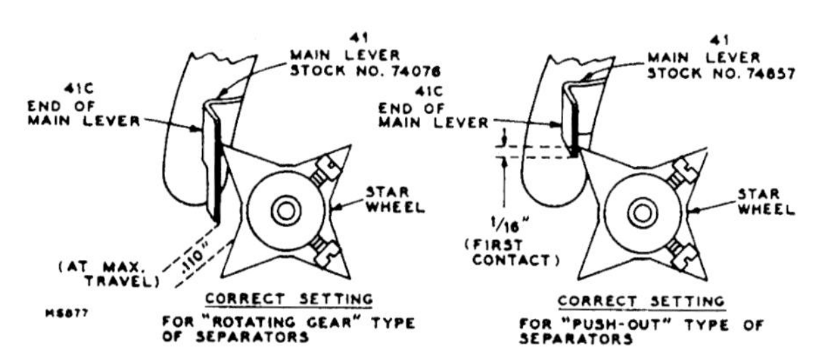

Figure 35 Setting of Director Lever.

Pivot Screw Adjustment:

Loosen the pivot locking screw (14) and adjust the pivot screw (12) for minimum side play without causing binding.

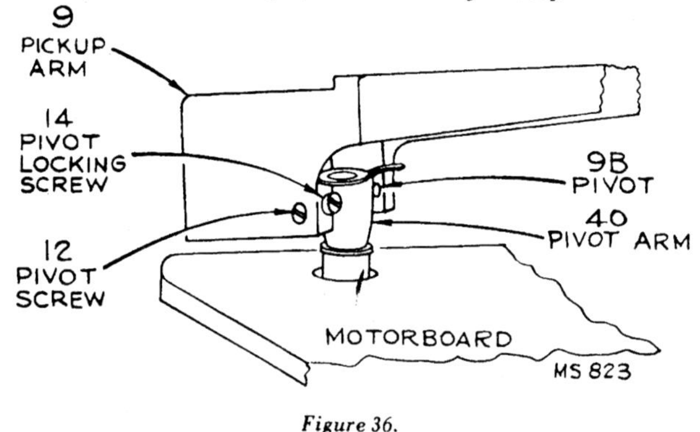

Figure 36.

Sapphire Height Adjustment (Out of Cycle):

Bend the lug on the pivot arm (40) so that the sapphire point is approximately 1/16" above the motorboard.

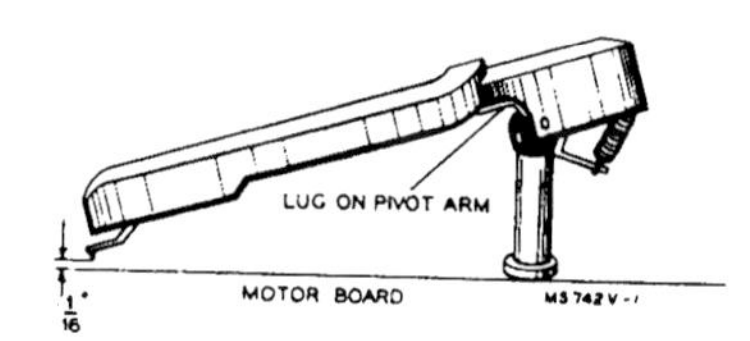

Figure 37.

Tripping Adjustment:

1. Assemble the pickup arm and trip lever assemblies as shown in Figure 38. Leave the clamping screw (57) loose enough to permit horizontal movement of the trip lever on the shaft. (Allow approximately .010 inch vertical end play.)
2. Turn the eccentric landing adjustment stud (45C) to determine the inward and outward limit of adjustment, then turn it to a setting half-way between the limits.

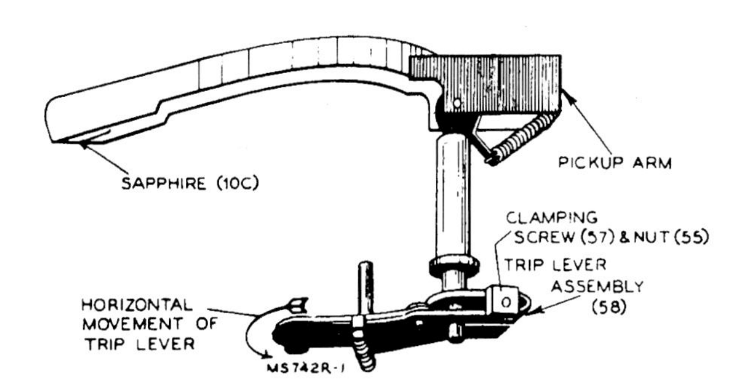

Figure 38.

3. Tripping should occur when the sapphire reaches a position 1 9/32" from the near side of the turntable spindle. This position is adjusted by holding the trip lever and moving the pickup arm inward or outward to obtain the specified position.
4. A convenient way of measuring this distance is to make a mark on the back side of a stroboscope disc 1 9/32" from the inner edge, place the disc on the turntable, with the turntable revolving, hold the disc stationary and move the pickup arm very slowly in towards the turntable spindle.
5. After this position has been obtained, tighten the clamping screw (57) and recheck the tripping position and vertical end play.

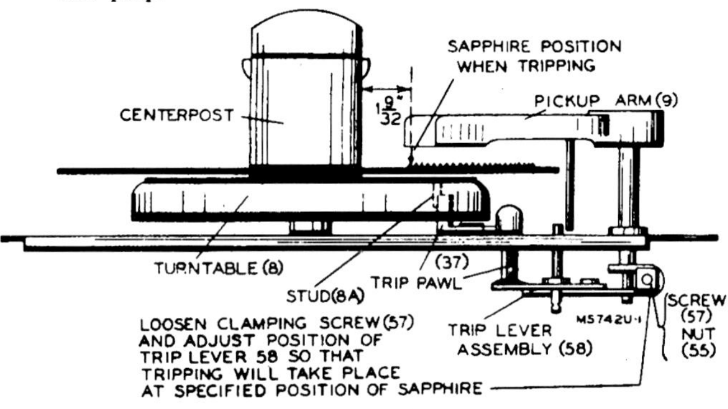

Figure 39—Tripping Position.

Landing Adjustment:

1. After the tripping adjustment has been made as described above, turn the eccentric landing adjustment stud (45C) so that the sapphire will set down on the record half-way between the outer edge and the first music groove. This position is 2⅝" from the turntable spindle. The location of the adjustment stud is illustrated in Figure 42.

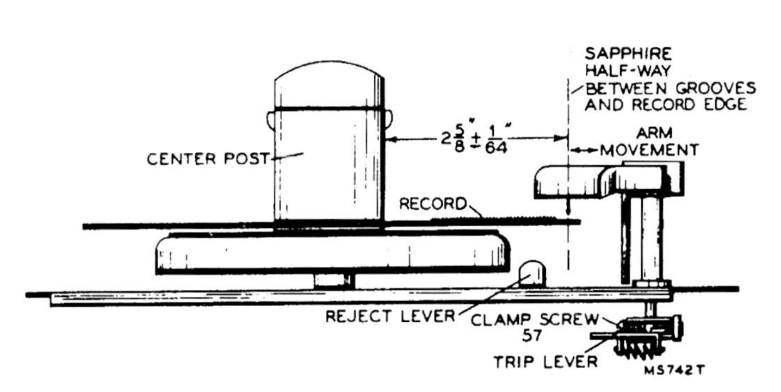

Figure 40—Landing Position.

Pickup Arm Height Adjustment (In Cycle):

Set the mechanism in cycle. Turn the turntable by hand, until the pickup arm has reached its maximum height. By means of a screwdriver turn the height adjustment stud (45D) until the distance between the top of the turntable and the sapphire point is ¾". Use that position of the eccentric stud which causes the pickup arm to rise during clockwise adjustment of the stud. The location of the adjusting stud is illustrated in Figure 42.

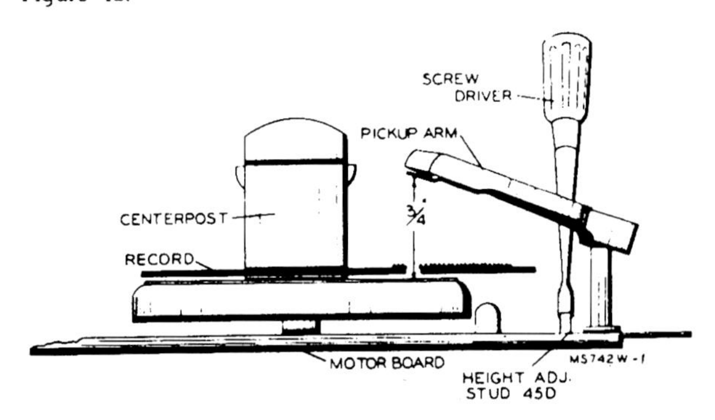

Figure 41—Height Adjustment.

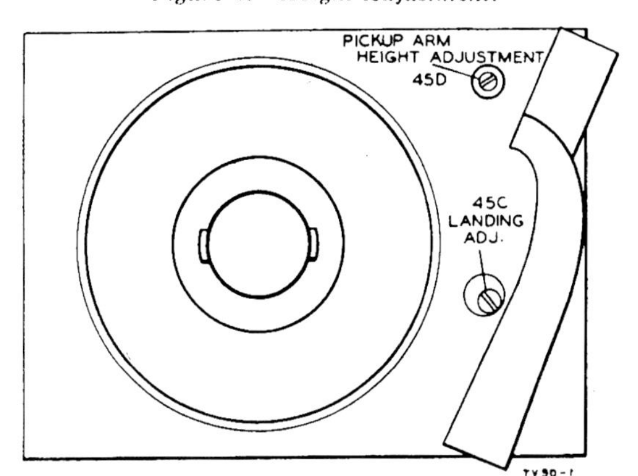

Figure 42—Height and Landing Adjustment Studs.

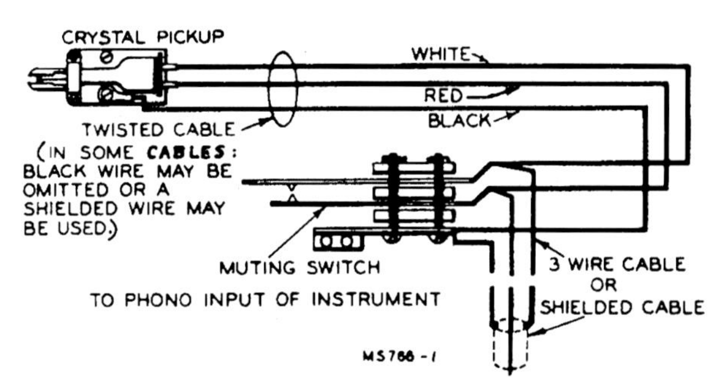

Figure 43—Pickup Muting Switch Wiring.

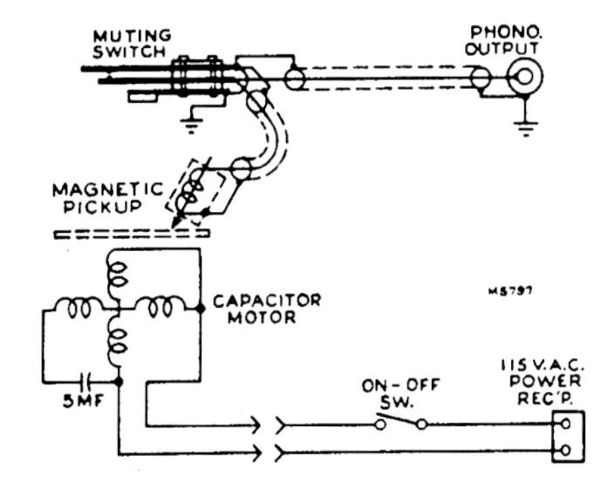

Figure 44—Schematic Diagram (Model CP-5203).

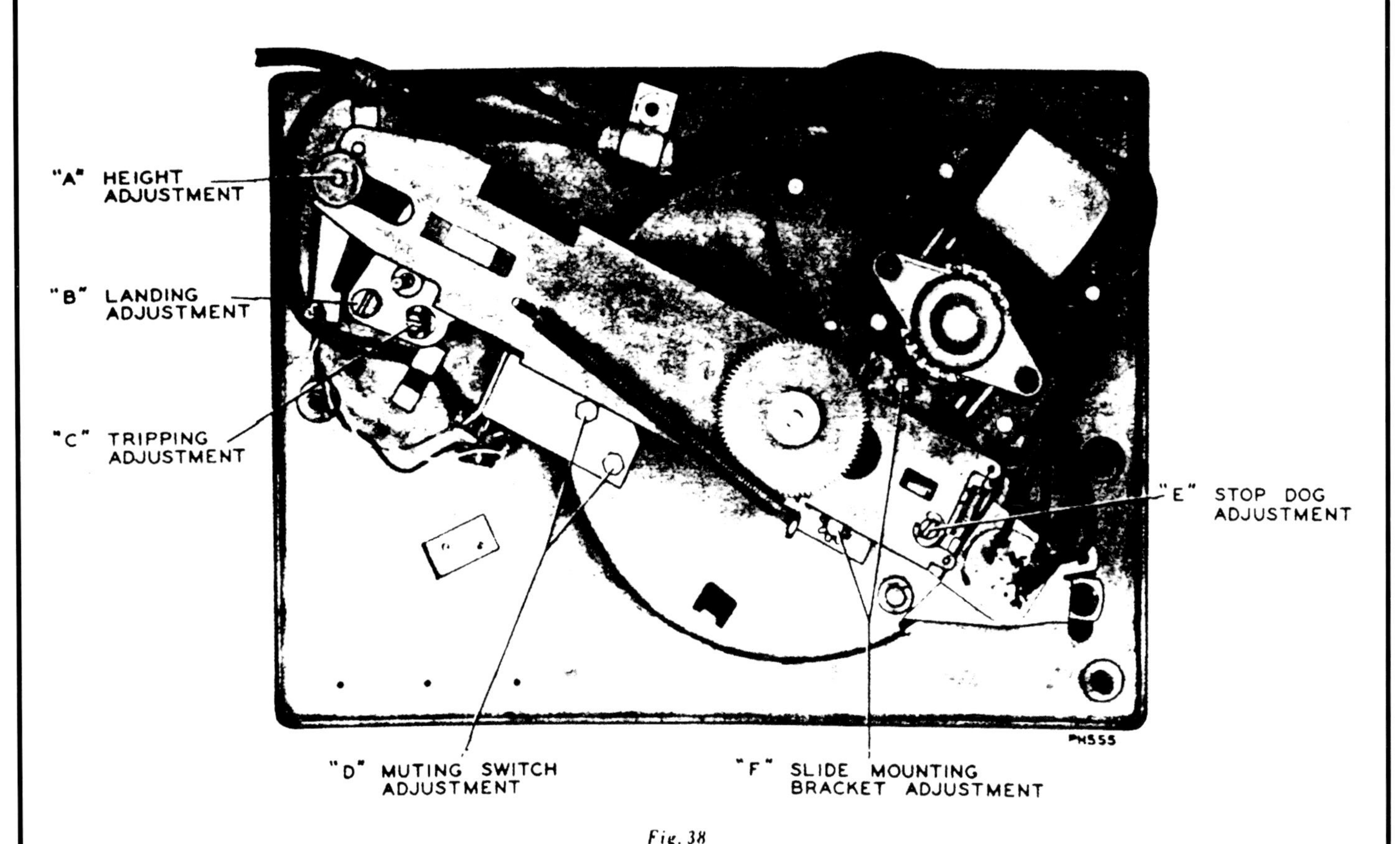

Fig. 38

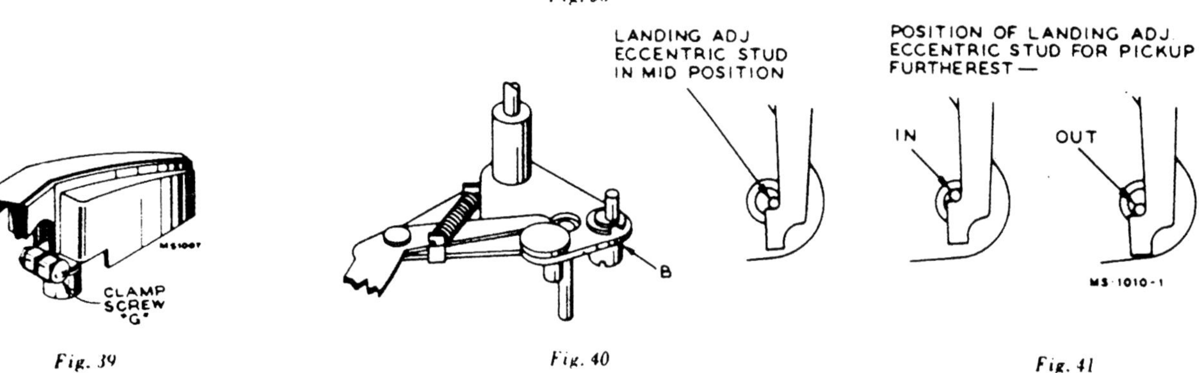

Fig. 39 *Fig. 40* *Fig. 41*

Adjustments

Pickup Landing Adjustment:

Under ordinary conditions the landing adjustment is a screwdriver adjustment as shown. The adjustment of eccentric landing adjustment stud (B) gives approximately a ¼" movement. (See Figs. 38, 40.)

If, however, the pickup arm has been removed it is first necessary to make an approximate landing adjustment as follows:

1. With the mechanism out of cycle and the clamp screw (G) (Fig. 39) loose, place pickup arm on the rest and tighten clamp screw enough to prevent the clamp from slipping on the shaft.
2. Set the landing adjustment stud (B) as shown (mid-adjustment). (See Figs. 40, 41.)
3. With the power removed, push reject control to reject. Rotate turntable by hand in the correct direction until the pickup is about ready to land.
4. Loosen clamp screw (G) and move pickup arm so the stylus is approximately 2⅜" from side of centerpost. Tighten clamp screw. (See Figs. 36, 39.)
5. Exact landing adjustment can now be made by a screwdriver on stud (B). (See Fig. 38.)

Pickup Height Adjustment (See Fig. 38):

Adjust knurled nut (A) until the distance (during change cycle) between the top of the turntable and the stylus point is approximately 1⅛".

NOTE: If unable to adjust for sufficient height, it may be necessary to cut a few turns from the compression spring to allow more space on the shaft.

Tripping Adjustment (See Figs. 37, 38):

Adjust the eccentric tripping stud (C) until the mechanism trips when the stylus is 1 9/32" from the side of the centerpost.

Mounting Bracket Adjustment (See Fig. 38):

Loosen the two screws (F) and move the bracket so it is as near perpendicular to the slide as possible. Move back or forward until the cut away section of the cycling cam clears the knurled roller approximately 1/16". Tighten screws.

Muting Switch Adjustment (See Fig. 38):

Loosen the two screws (D) and adjust the position of the switch so the contacts are approximately 1/32 to 1/16 inches apart when the mechanism is out of cycle. If the mounting screws do not give sufficient adjustment, bend tab on slide slightly.

Stop Dog Adjustment (See Fig. 38):

Turn the eccentric screw (E) until the record drops to the turntable without striking the pickup arm.

ADJUSTMENTS

PICKUP ARM HEIGHT

Loosen the screw marked ("A") on back of the pickup arm and adjust so the pickup will clear a stack of twelve records. Raising the screw in the elongated hole raises the pickup arm, lowering the screw lowers the pickup arm.

PICKUP·LANDING ADJUSTMENT

Loosen screw marked ("B") and slide the mounting bracket forward to move the landing point away from the centerpost, and back to move the landing point inward.

NOTE: Before making the adjustment, make certain the safety springs (26) are in the pin grooves.

TRIPPING ADJUSTMENT

If mechanism fails to trip when the stylus is approximately 1 9/32" from the side of the centerpost, bend the end of the segment engagement lever (indicated in drawing at right) out for early tripping and in for late tripping.

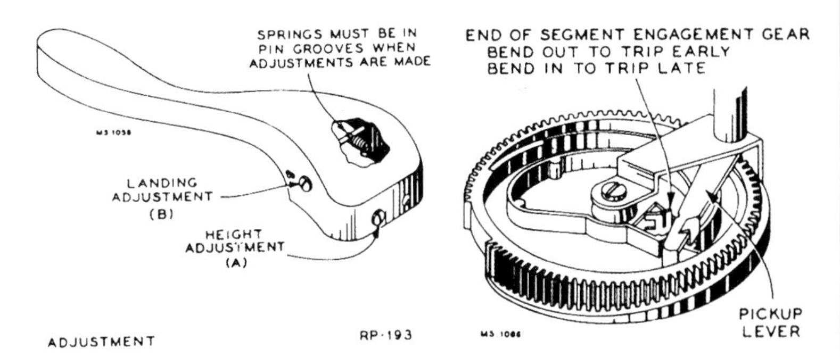

Note:
If spacing between separator blades and separator shelves do not fall between .040 to .048" bend blades accordingly.

RP-193

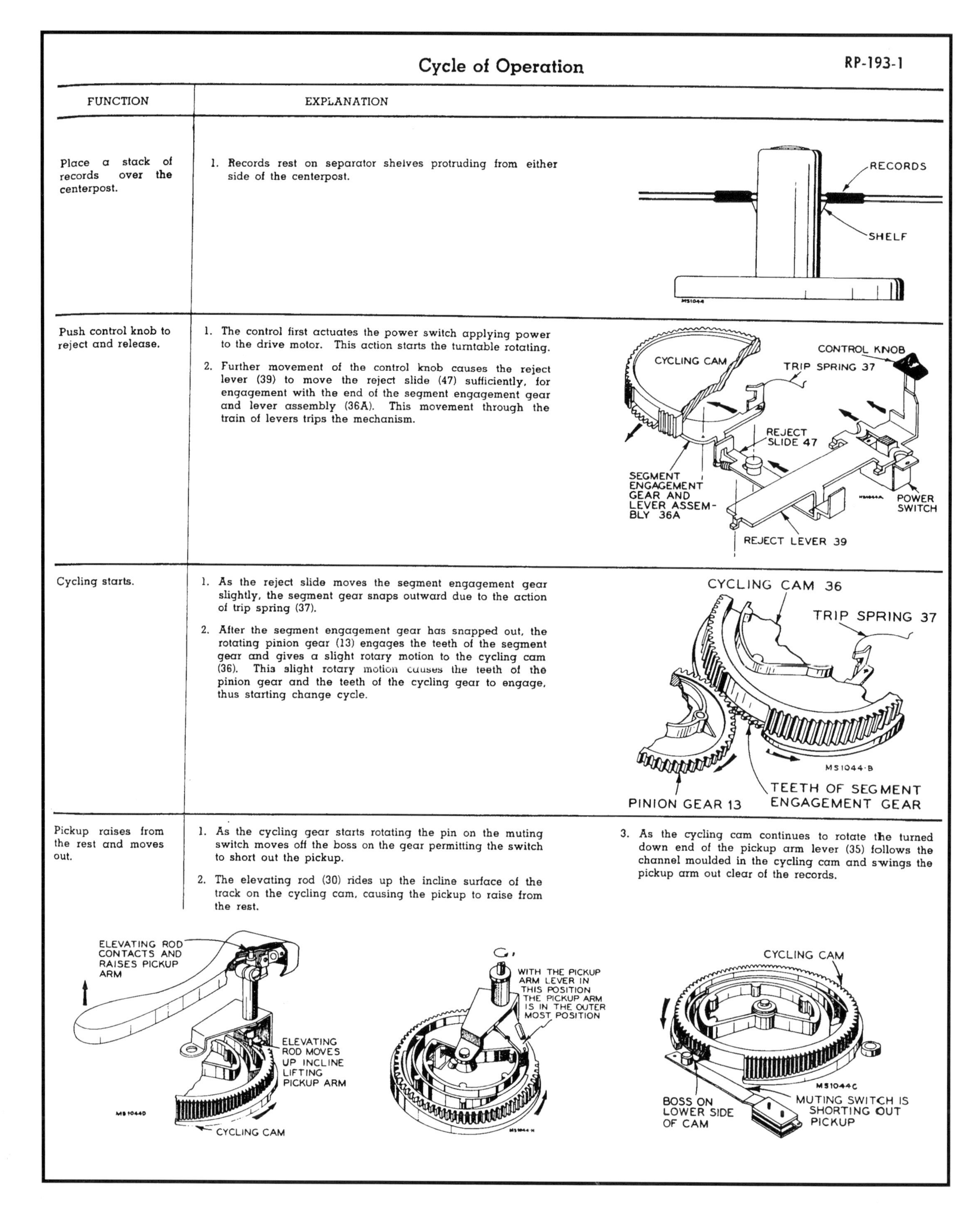

Cycle of Operation

RP-193-1

FUNCTION	EXPLANATION
Place a stack of records over the centerpost.	1. Records rest on separator shelves protruding from either side of the centerpost.
Push control knob to reject and release.	1. The control first actuates the power switch applying power to the drive motor. This action starts the turntable rotating. 2. Further movement of the control knob causes the reject lever (39) to move the reject slide (47) sufficiently, for engagement with the end of the segment engagement gear and lever assembly (36A). This movement through the train of levers trips the mechanism.
Cycling starts.	1. As the reject slide moves the segment engagement gear slightly, the segment gear snaps outward due to the action of trip spring (37). 2. After the segment engagement gear has snapped out, the rotating pinion gear (13) engages the teeth of the segment gear and gives a slight rotary motion to the cycling cam (36). This slight rotary motion causes the teeth of the pinion gear and the teeth of the cycling gear to engage, thus starting change cycle.
Pickup raises from the rest and moves out.	1. As the cycling gear starts rotating the pin on the muting switch moves off the boss on the gear permitting the switch to short out the pickup. 2. The elevating rod (30) rides up the incline surface of the track on the cycling cam, causing the pickup to raise from the rest. 3. As the cycling cam continues to rotate the turned down end of the pickup arm lever (35) follows the channel moulded in the cycling cam and swings the pickup arm out clear of the records.

RP-193-1

Separator blades separate the lower record from the stack and the lower record drops to the turntable.

1. An instant after the pickup arm has started to raise the rotating pinion gear (13) starts to raise also. This is due to the lower edge of the gear riding up the spiral incline formed on the edge of the cycling cam.

2. The raising of the pinion gear and key assembly actuates the separating mechanism inside the centerpost. This action causes the support shelves to recede and the separator blades to move out to select the lower record of the stack and to support the remaining records while the bottom record drops to the turntable.

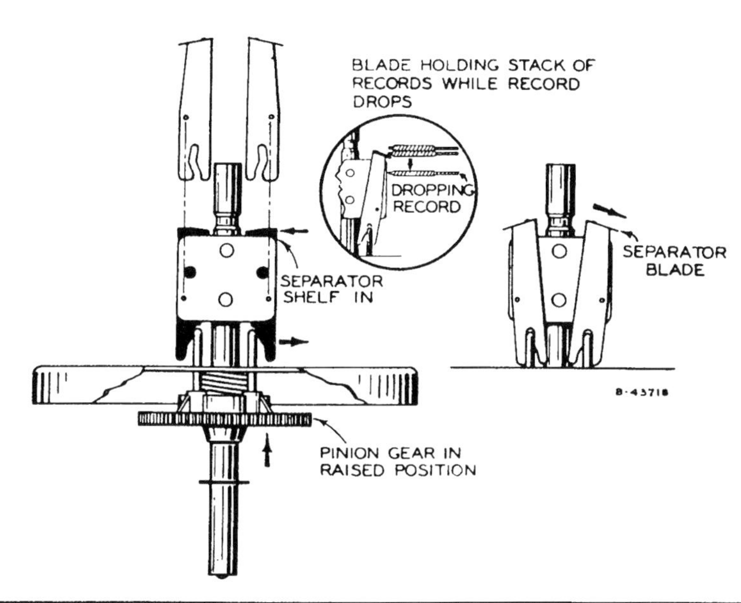

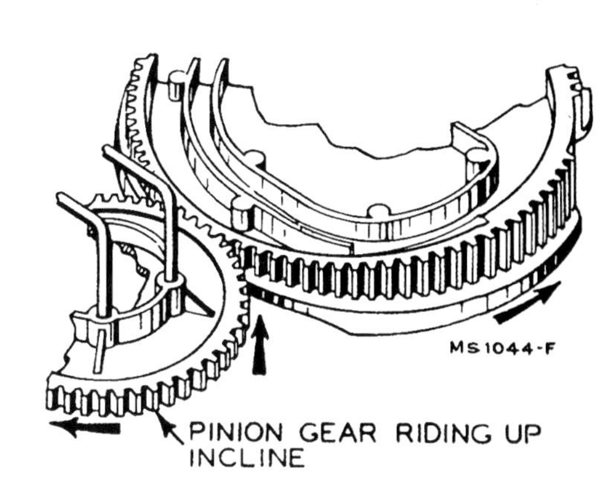

Pickup moves in for landing.

1. As the mechanism nears the end of the change cycle, the end of the segment engagement gear extending from the top of the rotating cycling gear comes against the mounting bracket. This contact resets the segment engagement gear preventing continuous cycling.

2. The end of the pickup arm lever riding in the channel in the cycling gear, moves the pickup arm in for landing.

3. The pickup lands on the start of the record as the elevating rod rides down the incline on the cycling gear.

4. At this very moment the end of the pickup arm lever moves into the open portion of the cycling cam track. This gives free movement to the pickup arm as it moves across the record.

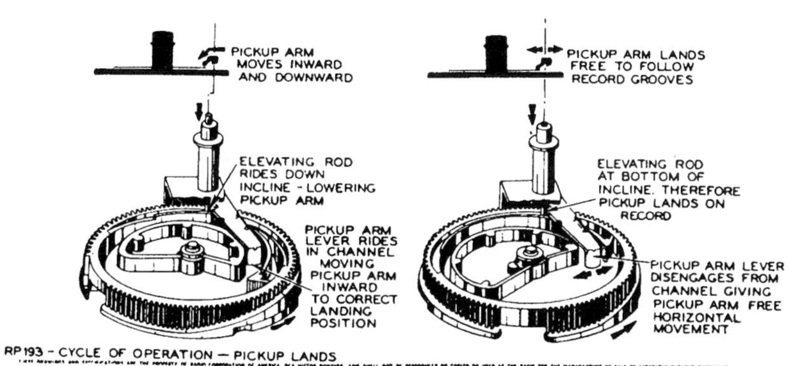

RP193 - CYCLE OF OPERATION — PICKUP LANDS

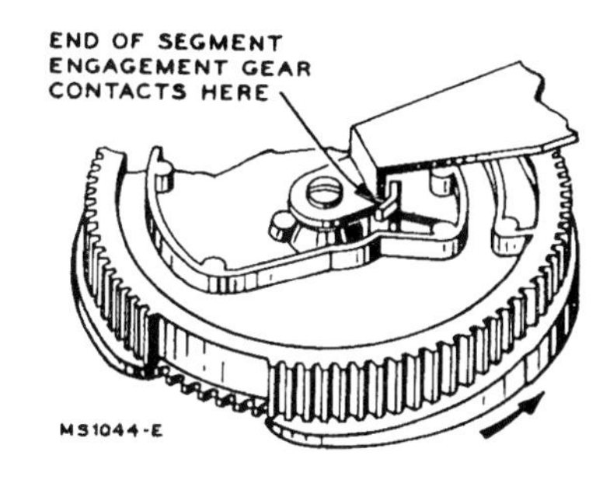

Cycling completed and the record plays.

1. The detent lever (17A) snaps the cycling cam into a neutral position as the muting switch pin comes in contact with the boss on the cam. This completes the change cycle.

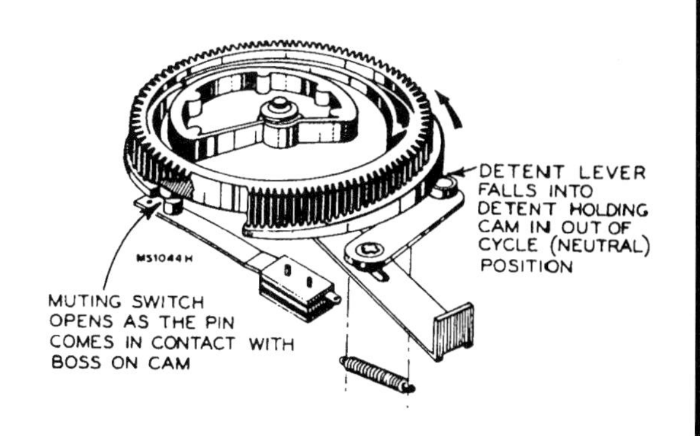

2. As the record plays, the pickup moves inward.

3. When the stylus reaches the end of the selection, the side of the pickup arm lever (35) contacts and trips the segment engagement gear and a new change cycle is started.

4. The mechanism repeats the preceding sequence of operations until the last record of the stack has dropped to the turntable and has been played.

5. The last record will be repeated until the pickup is lifted and placed on the rest.

RP-193-1

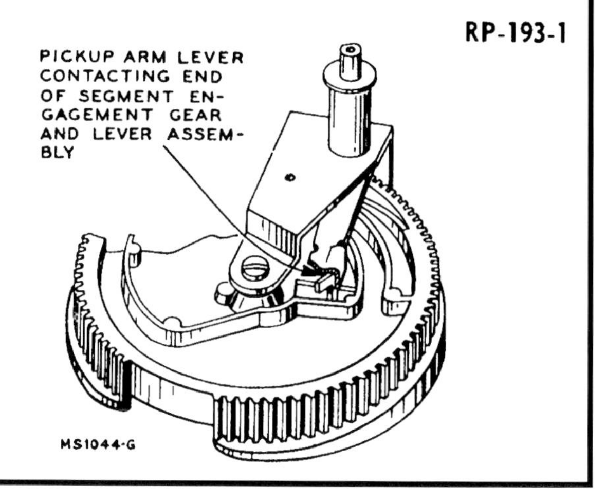

Appendix B

List of RCA Victor Table Model Phonographs

Listed here are all RCA Victor table model phonographs and radio/phonographs that were introduced from 1949 through 1957.

Model	Year	Changer	Amp	Tonearm	Record Player Description
9EY3	1949	rp-168	yes	gold	3 tube amp/brown Bakelite case
9EY31	1949	rp-168	yes	gold	3 tube amp/tan leatherette case
9EY32	1949	rp-168	yes	gold	3 tube amp/red leatherette case
9EY35	1949	rp-168	yes	gold	Disney motif/3 tube amp/white Bakelite case
9EY36	1949	rp-168	yes	gold	Roy Rogers motif/3 tube amp/white Bakelite case
9EY3M	1949	rp-168	yes	gold	3 tube amp/single play/brown Bakelite case
9JY	1949	rp-168	no	gold	brown Bakelite case.
9Y51	1949	rp-168	yes	gold	radio/phono with slide rule dial/5 tubes/Bakelite case with lid
9Y7	1949	rp-168	yes	gold	radio/phono with slide rule dial/6 tubes/wooden case with lid
45EY1	1950	rp-168	yes	maroon	3 tube amp/brown Bakelite case
45EY	1950	rp-168	yes	two-tone	3 tube amp/brown Bakelite case
45EY15	1950	rp-168	yes	gold	Disney motif/3 tube amp/white Bakelite case
45EY2	1950	rp-190	yes	black	3 tube amp/brown Bakelite case
45EY3	1950	rp-190	yes	black	3 tube amp/brown Bakelite case with lid
45J	1950	rp-168	no	gold	brown Bakelite case.
45J2	1950	rp-190	no	black	brown Bakelite case.
45J3	1950	rp-193	no	gold	brown Bakelite case.
9Y510	1950	rp-190	yes	black	radio/phono with slide rule dial/5 tubes/Bakelite case with lid
15E	1951	rp-190	yes	black	record demonstrator with two turntables/5 tubes/wooden case
45EY26	1951	rp-190	yes	red	Alice in Wonderland motif/3 tube amp/white Bakelite case
45EY4	1951	rp-190	yes	black	4 tube amp/brown Bakelite case with lid
9Y511	1951	rp-168	yes	gold	radio/phono with slide rule dial/5 tubes/Bakelite case with lid
45HY4	1954	rp-190	yes	black	4 tube amp/black Bakelite case with lid
4Y511	1954	rp-190	yes	black	radio/phono with round dial/5 tubes/black Bakelite case with lid
6BY4A	1955	-	yes	gray	portable radio/phono/5 tubes/black and gray plastic case/single play
6BY4B	1955	-	yes	pink	portable radio/phono/5 tubes/white and pink plastic case/single play
6EY1	1955	rp-190	yes	gray	3 tube amp/black Bakelite case
6EY15	1955	rp-190	yes	orange	Ding Dong School/3tube amp/black Bakelite case
6EY2	1955	rp-190	yes	gray	3 tube amp/black Bakelite case with lid
6EY3A	1955	rp-190	yes	gray	portable phono/2 or 3 tubes/two-tone brown vinyl covered case
6EY3B	1955	rp-190	yes	gray	portable phono/2 or 3 tubes/two-tone green vinyl covered case
6EY3C	1955	rp-190	yes	gray	portable phono/2 or 3 tubes/two-tone blue vinyl covered case
6XY5A	1955	rp-199	yes	-	radio/phono/5 tubes/black and gray plastic case/single play Slide-O-Matic
6XY5B	1955	rp-199	yes	-	radio/phono/5 tubes/white and turquoise plastic/single play Slide-O-Matic
6JM1,2	1955	rp-199	no	-	Plastic case/assorted colors/single play Slide-O-Matic
6JM25	1955	rp-199	no	-	Ding Dong School/white and red plastic case/single play Slide-O-Matic
6JY1	1955	rp-190	no	gray	black Bakelite case
6JY1A	1955	rp-190	no	white	black Bakelite case
6JY1B	1955	rp-190	no	white	white Bakelite case
6JY1C	1955	rp-190	no	white	green Bakelite case
7EY1DJ	1955	rp-190	yes	white	1 tube amp/black and gray plastic case
7EY1EF	1955	rp-190	yes	white	1 tube amp/pink and white plastic case
7EY1JF	1955	rp-190	yes	white	1 tube amp/coral and gray plastic case
7EY2HH	1955	rp-190	yes	peach	3 tube amp/two-tone green plastic case
7EY2JJ	1955	rp-190	yes	peach	3 tube amp/two-tone gray plastic case
7EP45	1956	rp-190	yes	gray	portable phono/3 tube amp/blue vinyl case with Elvis signature in gold
7HF45	1956	rp-190	yes	white	4 tube amp/wood case (mahogany, maple, or oak finish)
8HF45P	1956	rp-190	yes	white	portable phono/4 tube amp/two-tone brown vinyl covered case
8EY4DJ	1956	rp-190	yes	white	4 tube amp/dark Bakelite case with lid and gray front
8EY4FK	1956	rp-190	yes	white	4 tube amp/dark Bakelite case with lid and beige front
8EY31HE	1957	rp-190	yes	white	portable phono/1 tube amp/green and white vinyl case
8EY31KE	1957	rp-190	yes	white	portable phono/1 tube amp/rust and white vinyl case

Appendix C

Resources for Parts and Services

The following is a list of places where important parts and services can be obtained.

Antique Electronic Supply
6221 S. Maple Ave
Tempe, Arizona 85283
www.tubesandmore.com
480-820-5411
(sells tubes, capacitors, etc.)

Antique Radio Classified
P.O. Box 2
Carlisle, Massachusetts 01741
ARC@antiqueradio.com
978-371-0512
(magazine dedicated to the antique radio hobby)

New Jersey Antique Radio Club
c/o Phil Vourtsis
13 Cornell Place
Manalapan, New Jersey 07726
www.njarc.org
732-446-2427
(will answer questions related to 45 rpm system of recorded music)

Pembleton Electronics
1222 Progress Road
Fort Wayne, Indiana
219-484-1812
(sells motor grommets, capacitors, cartridges)

Radio Daze
7 Assembly Drive
Mendon, New York 14506
info@radiodaze.com
716-624-9755
(sells antique phonographs)

Waves
251-West 30th Street
New York, New York 10001
c1wave@aol.com
212-461-7121
(sells antique phonographs)

West Tech Services
P.O. Box 130
Prospect, Tennessee 38477
1-888-828-8455
(will rebuild cartridges and idler wheels)

Need a pro to restore your 45 phonograph? These folks specialize in such work.

Willie Bosco
1588 Miller Creek Rd.
Garberville, California 95542
williambosco@msn.com
707-923-3897
(sells new cycling cam rubber kit and cartridges)

Paul Childress
6632 W. Denny Ct.
Chesterfield, Virginia 23832
pchildress@prodigy.net
804-271-7842 pin 8707
(professional restoration)

Bob Havalack (Eager Bros. Music)
44 Backus Street (2nd Floor)
Rochester, New York 14608
ebm21@frontiernet.net
716-458-0801
(professional restoration)

Endnotes

[1]Republished with a different title with author's permission from Chapter 6 of dissertation, "Shaping the Sound of Music: The Origins of the LP and High Fidelity, 1939-1950." College Park, Maryland: Dissertation submitted to the Graduate School of the University of Maryland, 2000.

[2]See Warren Rex Isom, "Before the Fine Groove and Stereo Record and Other Innovations. . . ," *Journal of the Audio Engineering Society* 25, no. 10/11 (October/November 1977), 815.

[3]Goddard Lieberson, "A 33-1/3 Revolution in Recording," *SRL*, June 26, 1948, 40-41; Gelatt, Chapter 22, "Renaissance at a New Speed," 290ff.

[4]This account revises earlier histories of the phonograph industry. Gelatt, 290-96, and Oliver Read and Walter Welch's *From Tin Foil to Stereo: Evolution of the Phonograph*, 2nd ed. (Indianapolis: 1976), 333-42, focus on classical recordings and denigrate RCA's innovation. C. A. Schicke's *Revolution in Sound: A Biography of the Recording Industry* (Boston: 1974), 126-30, is dedicated to Wallerstein and also sees no logic in the 45. Andre Millard follows their lead in *America on Record: A History of Recorded Sound* (NY: 1995), 257.

[5]For industry sales, see Stephen Traiman, "The Record Industry in the United States," *Journal of the Audio Engineering Society* (hereafter *JAES*) 25, no. 10/11, 786; and Christopher H. Sterling and Timothy R. Haight, *The Mass Media: Aspen Institute Guide to Communication Industry Trends* (New York: 1978), Table 650-A, 356. For RCA's sales growth, which was higher than the industry's, see David Giovanonni, "The Phonograph as a Mass Entertainment Medium: Its Development, Adaptation, and Pervasiveness," (M.A. thesis, University of Wisconsin-Madison: 1980), table 15, 201; "Record Player Production from 1934 to 1942 Incl.," Historian's series VI, box 9, folder 45, Accession 2069: RCA/GE/Martin Marietta Collection, Hagley Museum and Library, Wilmington, Delaware (hereafter Hagley 2069); "Comparison Records vs. Instruments, Units, Victor, RCA and RCA Victor, 1901-1944," Historian's series VI, box 6, folder 46, idem.

[6]For Joyce and Wallerstein's work, see "Victor Record Society to Boost Sales," *The Scanner* 3, no. 4 (April 1938): 19, (Library of American Broadcasting, College Park, Maryland, hereafter LAB); "Enlarged General Record Catalogue Released" and "New RCA Air Show Stars Ace Bandmen," *The Scanner* 3, no. 5 (May 1938): 22 and 22-3 respectively; "Phonograph Records," *Fortune*, September 1939, 94; C. A. Schicke, *Revolution in Sound: A Biography of the Recording Industry* (Boston: 1974), 96-97.

[7]"Phonograph Records: Recovery in the Musical Reproduction Business," *The Index* (Liberty National Bank, New York), Autumn 1940, 68; "A Review of RCA Manufacturing Company's Radio Receiver and Phonograph Activities Yearly from 1933 to 1940, Inclusive, in Comparison with Sales by All Licensees," April 2, 1941, Hagley 2069, Historian's series VI, box 9, folder 49.

[8]"Automatic Record Changers: Prospect of Post-War Industry Production," July 11, 1944, pp. 1-2, Hagley 2069, Historian's series VI, box 9, folder 45.

[9]Oliver Read and Walter L. Welch, *From Tin Foil to Stereo: Evolution of the Phonograph*, 2d ed. (Indianapolis: 1976), 314-19; for more on David Rockola and the history of the Rockola Manufacturing Company, see William Bunch, *Jukebox America: Down Back Streets and Blue Highways in Search of the Country's Greatest Jukebox* (New York: 1994), 91-107.

[10]Roland Gelatt, *The Fabulous Phonograph: From Edison to Stereo*, 2d ed. (New York: 1965), 276, for the low estimates; "Phonograph Records: Recovery," 66, 69, for the higher numbers.

[11]My thanks to Susan Douglas, University of Michigan, for pointing out the link between disc jockeys using playlists and consumers buying more record changers; see Douglas, *Listening In: Radio and the American Imagination* (New York: 1999), 229, 246.

[12]"Phonograph Records: Recovery," 72.

[13]Warren Rex Isom, "Record Materials Part II: Evolution of the Disc Talking Machine," *JAES* 25, no. 10/11 (October/November 1977), 722; and Edward Tatnall Canby, "The New Recordings: How to Relax," *The Saturday Review of Literature* (hereafter *SRL*), August 10, 1946, 29.

[14]B. L. Aldridge to Henry G. Baker, February 11, 1946, Hagley 2069, Historian's series VI, box 4, folder 5; Benjamin Carson, Alexander Burt, and Hillel Reiskind, "A Record Changer and Record of Complementary Design," *RCA Review* 10, no. 2 (June 1949): 174. See also the comments on changers at the beginning of "Music for the Home," *Fortune*, October 1946, 156.

[15]"RCA Victor's Overall Automatic Record Changer Activity by Changer Mechanism," October 2, 1945, Historian's series, box 9, folder 45, Hagley 2069. See also Kurt List for record changer problems generally in "The Development of the Record Industry and its Techniques," in Neil F. Harrison, ed., *Record Retailing Yearbook* (New York: 1949), 8.

[16]James W. Murray, "Standardization and RCA," *SRL*, January 29, 1949, 49; "Case Report on RCA Victor Records," J. Walter Thompson Archives, Account files, box 16, folder: "RCA Account History (records through 1977 [?])," p. 45, John W. Hartman Center for Sales, Advertising, and Marketing History, Rare Book, Manuscript, and Special Collections Library, Duke University (hereafter Hartman Center). I have assumed that Joyce, as head of the Marketing Department for RCA Victor's Home Instruments, and Carson, who had designed every changer since 1927, were the principals involved. This is modeled on the routine followed in the Sound Engineering Division, although ideas could also come from other engineering departments or sources outside the company. See M. C. Batsel, "Sound Engineering Division," *The Scanner* 4, no. 12 (December 1939): 13-14, and Chart I on 15. This timing would explain the occasional reference in RCA promotional literature in 1949 to eleven years of development for the 45. It is possible that the innovation was initiated by E. N. Deacon, a former radio tube engineer, who became responsible for New Product Market Research and Analysis in January 1938: see "Announcement," *The Scanner*, 3, no. 2 (February 1938): 30. He was given broad powers to investigate prospects from "a neutral position, reporting to the General Manager or the President." I have found no other references to Deacon in any capacity.

[17]Record albums for manual players assumed that the listener turned over each disc before proceeding to the next: A-B-A-B. Albums for automatics assumed the listener turned over the stack of 78s after playing a series of sides: A-A-B-B. For the problem of storage, see "Magic Key commercial March 26, 1939," NBC Papers Central Files, Correspondence, box 71, folder 80: "RCA Manufacturing Company, Victor Division 1939," State Historical Society, Madison, Wisconsin (hereafter NBC Papers); and "Analysis of the Market for Model 'X' Record Changer," March 3, 1944, Hagley 2069, Historian's series VI, box 9, folder 45, p. 1. For the issue of consoles, see "Analysis of the Market for Model 'X'," p. 1; "Proceedings of the Annual Meeting of Stockholders," (Radio Corporation of America: 1944, 1949), 149, in "Minutes and Proceedings: Annual Meetings of Radio Corporation of America Stockholders, 1945-1951," 184, 187-88, in the David Sarnoff Library, Princeton, New Jersey (hereafter DSL). For the size of pre-war televisions, see Michael Ritchie, *Please Stand By: A Prehistory of Television* (Woodstock, NY: 1994), 63.

[18]H. Olson, "Subjective Response-Frequency Tests," report Z-2, p. 1, Hagley 2069, series XXII, box 143; "Music for the Home," 157-59; W. Rex Isom, "Record Materials Part II," *JAES* 25, 10/11, 723; "Loudness of Familiar Sounds: music and speech intensities based upon Bell Laboratories' survey," unnumbered page from *Radio Today*, [1938], in Hagley 2069, box 290, "Historical files: Decibels, Spectrums, etc." This file has since been reorganized in Historian's series II.

[19]Specifically, the percentage was 13.6170: "10" Bluebird and Black Label Compound 10-34," RCA Victor Division Record Department Standardizing Notice, Subject: Disc Records Compound, December 9, 1946, courtesy W. Rex Isom. At five dollars a pound, Vinylite for a four-ounce record amounted to $1.25: J. Davidson, "Petrochemical Survey: An Anecdotal Reminiscence," *Chemistry and Industry* 19, May 1956, 395. My thanks to Dr. Jeffrey Meikle, University of Texas, Austin, for this citation. RCA Victor's Red Seal classical records retailed for two dollars; material accounted for a fraction of the nineteen cents attributed to manufacturing costs and overhead. Those expenses were less than a penny for a popular disc: "Phonograph Records," *Fortune*, September 1939, 94.

[20]Sarnoff's philosophy runs throughout his pronouncements on the place of RCA in public and private. Internally, however, he made the importance and utility of monopoly more explicit. Compare the pamphlet of his speech to stockholders on May 7, 1935, "Television," DSL, and the edited selection in David Sarnoff, *Looking Ahead: The Papers of David Sarnoff* (New York: 1968), 97-99, with his "Memorandum on Television Policy," April 18, 1935, pp. 1, 2, and 5, attached to David Sarnoff to the Board of Directors, Radio Corporation of America, May 8, 1935, folder 360, box 41, Office of the Messrs. Rockefeller: Business Interests, Record Group III 2 c, Rockefeller Family Archives, Rockefeller Archive Center, Valhalla, New York. For the marketing of industrial leadership at RCA Victor, see B. L. Aldridge to Joe [B. Elliott], November 28, 1945, p. 2, Historian's series VI, box 9, folder 45, Hagley 2069, and the advertisements for RCA products in *Scientific American*, April, October, and December 1949. For RCA's interest in engineering standards, see H. I. Reiskind, "Groove Dimensions of Commercial Records," Engineering Memorandum EM-2409, February 10, 1942, Hagley 2069, series XI, box 48.

[21]For more on Sarnoff's business philosophy and strategies, see Margaret B. W. Graham, *RCA and the VideoDisc: The Business of Research* (New York: 1986), 40-2, and Robert Sobel, *RCA* (New York: 1986), 10-11, 92-183, passim, especially 101, 169-70. Neither author acknowledges the regulatory threat overhanging RCA's sanctioned monopoly of radio communications patents from the 1920s through 1958 as a constant influence on Sarnoff's strategy: see "RCA's Television," *Fortune*, September 1948, reprinted in "RCA: The Years 1930-1978 Reprinted in Selected Articles from *Fortune*," (New York?: 1979?), 30. It seems likely that Sarnoff justified RCA's monopoly to the government and RCA's long-term investment in television systems to his board of directors as a benefit to the public and a protection against antitrust action, respectively.

[22]For the name, see *The 50-Year Story of RCA Victor Records* (NY: 1953), DSL, 29. The secrecy which surrounded this project reveals itself in the fact that RCA did not file for patents on the changer and record until March 1949, and that no engineering memorandums or technical reports refer to it. See Carson's patent no. 2,634,135, applied for March 31, 1949, and col. 7 referring to record applications applied for the same month; and the surviving technical file series and the RCA Camden card catalog now held at the Hagley Museum and Library. Several engineers who worked at Camden in the 1940s have also attested to the isolation of the development group.

[23]Carson et al., "A Record Changer," 174-75; see also the Experimental Test Records file, May 25, 1942-June 23, 1943, BMG Archives, New York City. My thanks to Dr. Michael Biel, Morehead State University, for providing copies of this file.

[24]Carson et al., "A Record Changer," 175-78. The volatility of shellac prices upset manufacturing cost estimates; for example, the "orange fine" shellac used for records jumped from twenty to thirty-two cents a pound between January 1939 and January 1940. Vinyl copolymers sold in the late 1930s for about fifty cents a pound. See "Plastics Industry," *Chemical and Metallurgical Engineering* (hereafter *CME*), January 1937, 97; "Plastics in 1941," *CME*, February 1941, 99; and the shellac price series in *CME* for January of each year, 1929-1945.

[25]"Case Report on RCA Victor Records," JWT Archives, p. 2; Carson et al., "A Record Changer," 183-84.

[26]Carson et al., "A Record Changer," 183-84.

[27]"Analysis of the Market for Model 'X'" and "RCA Victor's Overall Automatic Record Changer Activity," p. 3, Hagley 2069.

[28]"Analysis of the Market for Model 'X'," p. 4.

[29]Ibid.

[30]Ibid., p. 5, 7. Apparently someone at RCA Victor showed a twelve-inch 78 playing twelve minutes on a side to the editors of one magazine in 1945: see [Peter Hugh Reed], "Columbia's Long-Playing Disc: Editorial Notes," *The American Record Guide* (hereafter *ARG*), July 1948, 329.

[31]Charles Hobbs to the writer, p. 2, August 12, 1996. Mr. Hobbs began work at RCA Victor in Camden, New Jersey, in July 1942.

[32]"Folsom Heads RCA Victor," *Radio Age*, January 1944, 22; quote from Sobel, 140-41, also 145. Folsom started as vice-president for RCA Victor in January 1944 and was promoted to executive vice-president in June 1945. Folsom is the subject of two fawning articles: Dickson Hartwell, "The Man Who Sells Takes the Helm," *Nation's Business*, November 1949, clipping in "Addresses . . .," vol. xxiii (1949), 100-104, DSL, and "The Two-Man Team that Runs RCA," *Business Week*, November 17, 1951, 112-22.

[33]Frank Folsom, "RCA Manufacturing Grows," *Radio Age* 3, no. 2 (October 1944): 12.

[34]John L. Mastran, "How RCA Organization is Planned to Meet Changing Needs," *RCA Engineer*, February/March 1956, 32-33; Sobel, 145; Frank M. Folsom, "Radio Sets in Production," *Radio Age* 5, no. 1 (October 1945): 18; Robert W. Wythes, "Victor Talking Machine Company: Cost Accounting, General Accounting, Treasury Dept.," typescript (August 1968), [p. 3], courtesy of Mr. Nicholas F. Pensiero, and since deposited at the Hagley Museum and Library as part of Collection 2138. Mastran was the M.B.A., hired in July 1943.

[35]See Graham, 45-46, for another perspective.

[36]"Case Report on RCA Victor Television Sets," Account files, box 16, folder: "RCA Account History (television)," c. 1948, 6-9; and "Case Report on RCA Victor Records," 7-8, 45-46, both in JWT Archives. The agency's ignorance of the 45 is inferred from the lack of references to the agency in the Home Instrument marketing documents in Hagley 2069, and the absence of comment on the system in the Case Report until 1948.

[37]Arthur Van Dyck, interview by Edwin Dunham, November 10, 1965, Oral History Transcript AT-125, p. 30, Broadcast Pioneers Library collection, LAB.

[38]Sanjek, 222; "A Platter for the Lion," *Time*, February 24, 1947, 95.

[39]"Proceedings of the Annual Meeting of Stockholders," (Radio Corporation of America: May 5, 1942), 14, DSL; obituaries for Thomas F. Joyce, New York *Post*, New York *Times*, September 10, 1966, n.p., William F. Hedges Collection, file 137A: "The RCA Story," LAB; Charles Hobbs to the writer, June 29, 1996, p. 3.

[40]J. W. Murray, "'Nipper' Listens In," *Radio Age* 4, no. 1 (October 1944): 39.

[41]Charles O'Connell, *The Other Side of the Record* (New York, 1947).

[42]Michael Gray, "A History of the Sound of Columbia Records," *The Absolute Sound*, no. 119 (August/September 1999): 31. Wallerstein is among those who claim he first encouraged Paley to buy the record company and hire him to run it: Edward Wallerstein as told to Ward Botsford, "Creating the LP Record," *High Fidelity*, April 1976, 57.

[43]Gray, "A History," 31; Michael Hobson, "The Classic Interview: The Birth of the LP," with George Avakian and Howard Scott, *The Audiofile*, http://www.classicrecs. com/newsletter/newsletter/newsletter.chm?Article=6, visited February 17, 1999; Wallerstein, "Creating the LP," 57.

[44]Frank B. Walker, "Phonograph Comes Back," *Radio Age* 1, no. 2 (January 1942): 24.

[45]While never made explicit, the circumstantial evidence and contemporary industry gossip link the two record companies to the promotions: "Battle Over Records," *Business Week*, February 24, 1940, 50-51.

[46]Ibid.; "Up, Up Go Sales, When Columbia Drops Drab Dress for Records," *Sales Management*, December 15, 1940, 20-22; C. B. Larrabee, "Record Prices Drop as Columbia Stages Sensational Revival," *Printers' Ink*, December 20, 1940, 9-12, 69-70; Keith Hutchison, "Everybody's Business," *The Nation*, January 4, 1941, 20. See also the ad proofs for Columbia's 1940 magazine advertisements in the D'Arcy Masius Benton & Bowles Archives, box 44, folder: "Columbia Records, 1940-41, 1946," Hartman Center. Standard histories emphasize Wallerstein's price cut without referring to RCA Victor's activities beforehand.

[47]Sanjek, 215-16; "Magic Brain," *Radio Age* 1, no. 1 (October 1941): 20.

[48]"Case Report on RCA Victor Records," p. 6. Hunter's basic claim was insufficiently novel for the patent office; see United States Court of Customs and Patent Appeals, Patent Appeal No. 5395, In Re James H. Hunter—Phonograph Record (October Term, 1947), Office of the Clerk, United States Court of Appeals For The Federal Circuit, Washington, D. C.

[49]Gray, "A History," 31.

[50]Bruce Anderson, Peter Hesbacher, K. Peter Etzkorn, and R. Serge Densioff, "Hit Record Trends, 1940-1977," *Journal of Communication* 30, no. 2 (Spring 1980): table 1, p. 33; "Phonograph Records," *Fortune*, September 1939, 100; "Case Report on RCA Victor Records," pp. 5-6, 103; Robert P. Dupin, "The Non-Price Competition in the Phonograph Record Industry at the Manufacturers' Level," (M.B.A. thesis: Wharton School, University of Pennsylvania, 1957), 52, University of Pennsylvania Libraries. RCA Victor's earlier market share is extrapolated from the information in the last three sources. See the Columbia advertisements in the D'Arcy Masius Benton & Bowles Archives, box 44, folder: "Columbia Records 1940 Masterworks, Consumer Magazines," Hartman Center. The reduced surface noise did not offset the label's "reputation for less-than-exceptional sound that dogged it throughout the 78 rpm era": Gray, "A History," 31.

[51]For more on the AFM strike, see James P. Kraft, *Stage to Studio: Musicians and the Sound Revolution, 1900-1950* (Baltimore: 1995), chapter 6. J. Walter Thompson's "Case Report on RCA Victor Records," asserts from agency research that Decca never threatened Victor's position, and that Victor's market share suffered from the independents much more after the war: pp. 5-6, 103. The difference may lie in measurement by records sold versus dollars grossed; Victor benefited in the latter area by the higher prices on Red Seal classical discs.

[52]Read and Welch, 333; Sanjek, 215; Gray, "A History," 31.

[53]Sanjek, 215, states that Victor's "warehouses bulged with imported German wax and shellac from India." Alice K. Kean, "Record Market Opens Door to Higher Production," *Domestic Commerce* (United States Bureau of Foreign and Domestic Commerce), March 1947, 52, reports industry stockpiling during 1941. Perhaps she was referring only to Victor. Net imports of shellac dipped in 1940 before jumping thirty-five percent in 1941. Victor's unit sales of records jumped nearly fifty percent in 1940 from the year before, and another fifty-eight percent in 1941. See the bar chart in "Shellac and Other Forms of Lac," chart issued July 10, 1942, War Production Board RG 179, box 1737, Policy Documentation File 537.1404: "Shellac - Supply and Requirements Summaries, National Archives at College Park, Maryland (hereafter NACP); and "Comparison: Records vs. Instruments (units), Victor, RCA, and RCA Victor, 1901-1944," Hagley 2069, Historian's series VI, box 9, folder 46.

[54]"Lac Bug vs. Jitter Bug," *Newsweek*, April 27, 1942, 59; Read and Welch, 333-34; "Informal Conference of Shellac Subcommittee of Protective and Technical Coatings Industry Advisory Committees," October 30, 1942, Policy Documentation File 537.1405; Walter Woodruff, "Staff Report: Shellac," December 2, 1942 [ten-page draft], p. 3, 10, Policy Documentation File 537.141; "Production Conference of Phonograph Record Manufacturers," February 7, 1944, Policy Documentation File 585.6305M, all in RG 179, NACP.

[55]Repeated heating of shellac condenses water out of its molecular structure, turning it from a thermoplastic to a thermosetting resin: John Delmonte, *Plastics in Engineering*, 3rd ed. (Cleveland: 1949), 91. For bagasse, see T. R. McElhinney, "Research at Valentine Sugars," *The Sugar Journal* 10, no. 2 (July 1947): 2-3; Will J. Gibbens, jr., "Bagasse as a Source of Plastics," *Gilmore's Lousiana Sugar Manual* (1940-41), xiii-xvi. My thanks to Carl J. Martin for providing these articles. Mr. Martin worked at Valentine Sugars Company and its plastics subsidiary, Valite Corporation, before joining RCA Victor Records during the war.

[56]Sanjek, 216. For the drawbacks of the shellac substitutes, see Leon S. Kaye, *The Production and Properties of Plastics* (Scranton, PA: 1947), 175-79.

[57]"Production Conference of Phonograph Record Manufacturers," February 7, 1944, RG 179, Documentation file 585.6305M: "Phonograph Records - Industry Advisory Committee Meetings," NACP.

[58]Sanjek, 219.

[59]Frank M. Folsom, "Radio Sets in Production," 18; "Radio-Phonograph Combinations: Table Models vs Consoles," folder 49: "Sales Comparisons, 1934-1946" and "Automatic Record Changers: Prospects of Post-War Industry Production," both in Hagley 2069, Historian's series VI, box 9, folder 45.

[60]Murray, "'Nipper' Listens In," 39.

[61]E. W. Butler, "Lower Distribution Costs Sought," *Radio Age* 1, no. 4 (July 1942): 10-11; quote from Murray, "'Nipper' Listens in," 39.

[62]B. L. A. [Benjamin L. Aldridge] to Joe [Joseph B. Elliott], November 28, 1945, Hagley 2069, Historian's series VI, box 9, folder 45.

[63]Virtually all histories of the development of FM rest on Lawrence Lessing's *Man of High Fidelity* (New York: 1956), a biography of E. H. Armstrong sponsored by the Armstrong Foundation. Gary Frost, University of North Carolina, Chapel Hill, presented a paper, "Corporate Culture and High Fidelity: RCA and FM Radio, 1930-1940," at the Society for the History of Technology meeting, Detroit, October 9, 1999, in which he documented RCA's experimentation with FM transmissions in 1931 and of Armstrong's relationships with the RCA engineers. Frost is preparing this paper for publication in a future issue of *Technology and History*.

[64]"RCA Outlines FM Policies," *Radio Age* 3, no. 3 (April 1944): 23-24.

[65]RCA Laboratories, "FM Broadcasting," c. February-March 1944, DSL; "New Circuit Lowers Cost of FM Radios," *Radio Age* 5, no. 1 (October 1945): 23.

[66]G. L. Beers and C. M. Sinnett, "Some Recent Developments in Record Reproducing Systems," *Proceedings of the I. R. E.* (hereafter *PIRE*) 31, no. 4 (April 1943): 138-46, quotes on 138, 145; Nathaniel Rhita, "An FM Phono Pickup," *Radio-Craft*, November 1945, 106. Olson abandoned the "rather complicated and expensive" FM pickup for a diode electronic pickup, which had its own R&D challenges: quote from "Electronic Phonograph Pickup," p. 1, December 8, 1944, Harry F. Olson Papers, folder: "Pickup—Electronic, 1944-47," Sarnoff Corporation Library, Princeton, New Jersey (hereafter Olson Papers). See also Olson, "Mechano-Electric Transducers," *Journal of the Acoustical Society of America* (hereafter *JASA*) 19, no. 2 (March 1947): 307-19.

[67]Quotes from "RCA Outlines FM Policies," 23-4; O. B. Hanson, "Down to Earth on 'High Fidelity'," *Radio*, October 1944; McMurdo Silver, "High Fidelity," *Radio-Craft*, March 1945, 380-81.

[68]Douglas, *Listening In*, 262-63; see John M. Conly, "They Shall Have Music: FM to the Rescue!" *Atlantic Monthly*, January 1951, 91-

95, for a detailed review of the postwar growth in FM stations, and Conly, "They Shall Have Music: Tuners, Aerials, and FM," *Atlantic Monthly*, September 1955, 94. Douglas cites Lessing for the pre-war total of 500,000 FM receivers; RCA estimated that there were 350,000-500,000: "FM Broadcasting," 37.

[69]See, for example, the comments on the future in "Phonograph Records: Recovery in the Music Business," 71-72, which ignores magnetic media in favor of photoelectric cells reading film and phonograph pickups.

[70]See J. Guy Woodward's RCA Laboratories notebook no. 25 for the fall of 1943, new file case #1-5, Sarnoff Corporation, Princeton; Sanjek, 221; John T. Mullin, "Magnetic Recording for Original Recordings," *JAES* 25, no. 10/11 (October/November 1977), 698; and for the most thorough treatment, Mark Henry Clark, "The Magnetic Recording Industry, 1878-1960: An International Study in Business and Technological History," (Ph.D. diss., University of Delaware, 1992), condensed pagination version, especially Chapters 6 and 7 on the work at the Brush Development Company and the Armour Research Foundation, respectively.

[71]"Magnetic Recorders," *Radio-Craft*, July 1948, 35, based on James E. Jump, *SRL*, June, 1948; "Recording Sound on Wire," *Radio Age* 7 no. 2 (January 1948): 18, 31; "Magnetic Tape Recording," *Fortune*, January 1951, 97-106; Clark, "The Magnetic Recording Industry," 134.

[72]"Magnetic Recorders;" Sanjek, 226; "Editorial: Scorpion," *Radio*, April 1943, 13. I have seen no reference to RCA Victor's wire recorder beyond the announcement in *Radio Age*, n. 67 above.

[73]See Folsom's summary of RCA Victor's struggles in "Proceedings of the Annual Meeting of Shareholders," (1946), 72-73, DSL. The WPB fixed the price of shellac at thirty-six cents per pound for the duration. Although the price had more than doubled without adjustment for inflation, demand for vinyl copolymers tripled supply in 1946, even as producers struggled to produce enough plasticizers. See shellac price series in January issues of *CMF*, 1941-1946; F. H. Carman, "Plastic Materials: Review of Supply Outlook," *CME*, October 1946, 158-59. Kean, "Record Market," 54, states a high of seventy-three cents a pound.

[74]Traiman, "The Record Industry," 786. These figures are based on estimates made by the Recording Industry Association of America, which was not organized until 1951.

[75] "New Record Plant at Canonsburg, Pa.," *Radio Age* 6, no. 2 (January 1947): 19.

[76]"Case Report on RCA Victor Records," JWT Archives, 29, 41, 103 (Exhibit F). J. Walter Thompson's figures are drawn from Victor on the one hand and a special survey of consumers commissioned by the agency to analyze market preferences. Neither of these mesh with those given by the RIAA in the preceding note, and they conflict with each other on Victor's share of the market. In addition, JWT's numbers indicate that Victor's sales rose only a million dollars in 1946 from the 1945 total of $15,000,000. For an example of the origins of the independents, see "Smaller Record Firms Spin to Success," *Business Week*, September 7, 1957, 54. Charlie Gillett, in *The Sound of the City* (New York: 1970), surveys the growth of the rock-oriented independents, which also included folk, jazz, classical, and children's labels.

[77]"Proposed Standards of the Radio Manufacturers Association," *PIRE* 34, no. 4 (April 1946): 200W. Presumably for political reasons, the RMA favored the eccentric stopping groove initiated by Eldridge Johnson and maintained by RCA Victor, but those stamping records with a semicircular or a V-shaped edge could keep their preferences within certain bounds. This became the basis for increasingly international standardization in the record industry: see J. A. Caffiaux, "A Brief Review of EIA Standards in the Audio Field," *JAES* 16, no. 1 (January 1968): 212-5.

[78]R. C. Moyer, "Standard Disc Recording Characteristic," *RCA Engineer*, October-November 1957, 12, reprinted in Roys, ed., "Record Engineering," 10-11. The story of the development of a general recording standard is described in the articles and announcements concerning the development of the "Sapphire Group" and then the Audio Engineering Society in *Audio Engineering* between 1947 and 1951. Susan Schmidt Horning, Case Western Reserve University, is writing her dissertation on the evolution of sound recording studios and will cover this subject in detail.

[79]"Your Victrola's jewel-point pickup floats like a feather on water," advertisement on back cover of *Radio Age*, July 1946; I. Queen, "Postwar Features of Phonograph Pickups," *Radio-Craft*, September 1947, 36-37, 55; "On Pickups," *ARG*, December 1947, 97; Richard H. Dorf, "Checking Performance of Phono Pickups," *Radio-Craft*, August 1948, 33-34; I. Queen, "New Magnetic Pickups," *Radio-Electronics*, October 1948, 55; "Modern Crystal Phono Pickups," *Radio-Electronics*, 29; for the accented bass response of the crystal pickup, see [Peter Hugh Reed], "Editorial Notes," *ARG*, December 1948, 98. Gernsback changed the magazine's name in October 1948. For a thoroughly documented survey of pickup technologies, see B. [Benjamin] B. Bauer, "The High-Fidelity Phonograph Transducer," *JAES* 25, no. 10/11 (October/November 1977), 732-39.

[80]L. C. Harlow, "Mechanization of Record Pressing," in *Record Engineering at RCA* (Indianapolis: c. 1957), 18-19. This booklet is a compilation of articles from *RCA Engineer* published in 1955 and 1956. Copy in the writer's possession, courtesy of Carl Martin.

[81]L. G. Harlow, "Even Thickness Records," Engineering Memorandum EM-6000, October 14, 1946; L. A. Wood, "Physical Tests on Competitive Ten Inch Records," EM-6009, September 23, 1958, both in Hagley 2069, series XIII: "Indianapolis engineering memoranda."

[82]Frank M. Folsom, "Brand Names are Trusted," *Radio Age* 8, no. 3 (April 1948): 22. The EMs in the 6000 series are littered with comments on the need to maintain quality in the search for cheaper materials and processes.

[83]E. W. Butler, "Lower Distribution Costs Sought," *Radio Age* 2, no. 4 (July 1942): 11.

[84]L. C. Harlow, "Even Thickness Records," Engineering Memorandum EM-6000, October 14, 1946, p. 4; Leon Parker, "Vinylite (VYHH) Potentialities as a Record Material," EM-6010, August 16, 1948, p. 4, both in Hagley 2069, series XIII.

[85]Kean, "Record Market," 55.

[86] "Record for the Ages," *Newsweek*, September 3, 1945, 72; "Unbreakable Records," *Radio Age* 5, no. 1 (October 1945): 22; F. B. Stanley, "Vinyl Records," *Modern Plastics*, June 1948, 125-28. Reiskind received an RCA Victor Award of Merit for his vinyl composition: "1946 RCA Victor Award of Merit," Hagley 2069, Historian's series, box 2, folder 8. The association of red surfaces with classical music continued. Vox rerecorded a black plastic pressing of Mozart's "Eine Kleine Nachtmusik," on red plastic: "Recent Recordings," *Audio Engineering*, April 1948, 42; and the Concert Hall Society moved from red Vinylite 78s to LPs: "New Long-Playing Discs, *ARG*, June 1949, 319-20.

[87]"The Record Year," *Newsweek*, December 24, 1945, 102-104, quote on 104; "Wear of Phonograph Records," *Consumers' Research Bulletin*, May 1947, 30; Read and Welch, 339, date the discs' debut as October 1946; Sanjek, 226.

[88]Leon Parker, "Low Noise Formvar-Vinyl Acetate Record Compositions," EM-6007, July 5, 1948, p. 4-5; Parker, "Vinylite (VYHH) Potentialities as a Record Material," EM-6010, August 16, 1948, p. 4, both in Hagley 2069, series XIII.

[89]Edward Tatnall Canby, "The New Recordings," *SRL*, April 13, 1946, 92; H. F. Olson, "Preliminary Program Phonograph Record Research," (witnessed May 3, 1948), p. 5, in Herbert Belar's RCA Laboratories notebook P-4613, Sarnoff Corporation Library.

[90]The writer's mother recalls obtaining a second-hand, windup Victrola with her roommates for their senior year at St. Agnes School in Alexandria, Virginia, in the fall of 1947.

[91]The search for substitutes extended at least until 1950: G. P. Humfeld, "Production Run on Zein Record Compound," EM-6008, March 14, 1950, Hagley 2069, series XIII.

[92]R. K. Lashof, "Particle Size of Fillers," EM-6004, January 10, 1948, pp. 9-11; E. B. Blay, "The Use of Ethyl Cellulose in Low Noise Solid Records," EM-6005, March 12, 1948; Leon Parker, "Low Noise Formvar-Vinyl Acetate Record Compositions," p. 3, all in Hagley 2069, series XIII.

[93]James D. Parker, "High-Fidelity Broadcasting," *Radio-Craft*, August 1939, 73, 107-109. Parker was a CBS engineer. For the promotion of FM, see "At Long Last—Static-Free Radio!" *Radio-Craft*, April 1939, 588, 618-19, and Hugo Gernsback's editorials, "Whither Radio?" and "'Static-Less' Radio," *Radio-Craft*, August and December 1939. Gary Frost, University of North Carolina, Chapel Hill, believes that Armstrong wrote the article.

[94]Harry Paro, "Modernize Old Phonographs!" *Radio-Craft*, August 1939, 89, 108-109; advertisement for RCA Victor recorder in *Radio-Craft*, August and September 1939, 107 and 131, respectively.

[95]Quote from S. Ruttenberg, "Fidelity vs. Harmonics," *Radio-Craft*, May 1937, 664; A. C. Shaney, "Intermodulation—And it Relation to Distortion," *Radio-Craft*, May 1939, 652-53, 689.

[96]See Leo Fenway, "Sound is the Spur," *Radio-Craft*, June 1941, 738-39, to see how two radio servicemen in New York City branched into sound installations without reference to high fidelity, and articles on home recording in *Radio-Craft*, 1940-1, passim.

[97]Harry F. Olson, "Extending the Range of Acoustic Reproducers," *Proceedings of the Radio Club of America* 18, no. 1 (January 1941), 1, 9.

[98]John H. Potts, "Audio Aspects of Postwar Radio Engineering," *PIRE* 35, no. 12 (December 1947): 1404; [H. E. Muhleman], "Editorial: Post-War," *Radio*, June 1942, 4. Given publishing schedules, Muhleman probably wrote this well before the outcome at Midway was announced that month.

[99]Robert Teitelman, *Profits of Science: The American Marriage of Business and Technology* (New York: 1994), 62; John G. Frayne and Halley Wolfe, *Elements of Sound Recording*, (New York: 1949), v.

[100]See the capsule autobiographies under these men's names in "Audio Pioneers of the People and by the People," *Audio*, May 1962, 44-5, 48-50, 54-6; W. W. Wetzel, magnetic-tape research director for Minnesota Mining and Manufacturing Company after the war, also moved into high-fidelity development as a result of wartime employment: John M. Conly, "They Shall Have Music: Tape Recording," *Atlantic Monthly*, February 1954, 92.

[101]John M. Conly, "They Shall Have Music: Read All About It . . . ," *Atlantic Monthly*, February 1952, 90.

[102]Harold Lawrence, "About Music: 15 Years Later," *Audio*, May 1962, 84; Gelatt, 282-83; Read and Welch, 338.

[103]Vin Zelluff, ed., "Tubes at Work: High-Fidelity Phonograph," *Electronics*, December 1946, 178.

[104]The most extensive source is Irving Kolodin, "Turntable Talk with Arthur Haddy," *SRL*, July 28, 1951, 35-36, 50, quote on 36. See also Roland Gelatt, "The Other Side," *SRL*, October 25, 1947, 61; "Decca—The Upstart Reaches 70," http://homepages.enterprise.net/beulah/ffrr/decca.html, visited January 25, 2000; A. C. W. Haddy, "'ffrr' Lateral Feedback Recorder," in H. E. Roys, ed., *Disc Recording and Reproduction*, (Stroudsburg, Pennsylvania: 1978), 65; Gelatt, 282; and Sanjek, 227. For more on Decca's recording techniques that accompanied the technological developments, see Michael Gray, "From the Golden Age: The Birth of Decca/London Stereo," *The Absolute Sound* 43 (1985), 103-110; and Gray and Robert Moon, *Full Frequency Stereophonic Sound* (San Francisco: 1990). Decca's role in closing out the Battle of the Atlantic is not mentioned in Willem Hackman, *Seek & Strike: Sonar, Anti-Submarine Warfare and the Royal Navy 1914-1954*, (London: 1984). My thanks to Dr. Jon T. Sumida, University of Maryland, College Park, for suggesting this reference.

[105]"Music for the Home," *Fortune*, October 1946, 156-60, 190-95.

[106]Ibid., 156-60.

[107]Ibid., 190, quote on 161.

[108]Ibid., 161, quotes on 190.

[109]Edward Tatnall Canby, "Record Revue: How I Fell Into Audio," *Audio Engineering*, March 1952, 38-41, quotes on 38; and "Record Revue: I Fall Further into Audio," *Audio Engineering*, April 1952, 36-38. Canby added further recollections in "Audio Etc.: Radio Recall," *Audio*, June 1994, 14-18.

[110]Canby first referred to the "record buying public" in "The New Recordings," *SRL*, February 6, 1946, 26, and to the "record buyer" in "The New Recordings," *SRL*, April 13, 1946, 92. In this trend he followed that of the leading record magazine, which changed from *The American Music Lover* to *The American Record Guide* before the war.

[111]See Canby's column in *SRL*, February 6, 1946, 41, for his definitions of these categories.

[112]Edward Tatnall Canby, "The New Recordings," *SRL*, February 6, 1946, 26.

[113][Edward Tatnall Canby], "The New Recordings," *SRL*, January 19, 1946, 43.

[114]Edward Tatnall Canby, "The New Recordings: Don't Buy It!" *SRL*, December 7, 1946, 99.

[115][John H. Potts], "Transients: Introducing 'Audio Engineering,'" *Audio Engineering*, May 1947, 4.

[116][John H. Potts], "Editor's Report: Progress Report," *Audio Engineering*, September 1947, 4; Benjamin F. Tillotson, "Musical Acoustics," *Audio Engineering*, June-December 1947.

[117][Potts], "Transients: Introducing 'Audio Engineering,'" 4.

[118]Edward Tatnall Canby, "Record Revue," *Audio Engineering*, June 1947, 33.

[119]Edward Tatnall Canby, "Record Revue," *Audio Engineering*, October 1947, 29.

[120]Edward Tatnall Canby, "Some Highs and Lows," *SRL*, October 25, 1947, 62.

[121]Edward Tatnall Canby, "Record Revue," *Audio Engineering*, November 1947, 34.

[122]Edward Tatnall Canby, "Some Highs and Lows: How High is Fidelity?" *SRL*, November 29, 1947, 63.

[123]G. M. Nixon, C. A. Rackney, and O. B. Hanson, "Down to Earth on 'High Fidelity,'" March 27, 1944, reprinted in "Reports on Standards and Frequency Allocations for Postwar FM Broadcasting," Radio Planning Board, Panel 5, June 1944, copy in E. H. Armstrong Papers, box 159, Columbia University Special Collections, quotes from emailed copy courtesy of Gary Frost.

[124]McMurdo Silver, "High Fidelity," *Radio-Craft*, March 1945, 347, 380-82; Emerick Toth, "High Fidelity Reproduction of Music," *Electronics*, June 1947, 113. Silver, a "high-fidelity sound pioneer," was a regular contributor to the magazine; Toth wrote from the Radio Division of the Naval Research Laboratory. Hanson's article was reprinted in *Radio*, October 1944.

[125]John G. Goodell and B. M. H. Michel, "Auditory Perception," *Electronics*, July 1946, 142.

[126]First quote from Lawrence V. Wells to the editor, *PIRE* 35, no. 4 (April 1947), 378; second quote from Goodell and Michel, "Auditory Perception," 142; third from Harold Burris-Meyer, "The Place of Acoustics in the Future of Music," *JASA* 19, no. 4 (July 1947), 534. See also Benjamin F. Tillson's comments on "Musical Education" in "Musical Acoustics: Part IV," *Audio Engineering*, September 1947, 30-31.

[127]CBS commissioned a survey in June 1946 on the record market for its company-owned stations that asked no questions about sound quality: Surveys Division, CBS Research Department, "Selected Data on Musical Preferences and the Market for Records" May 1947, Performing Arts Division, Library of Congress.

[128]Howard A. Chinn and Philip Eisenberg, "Tonal-Range and Sound-Intensity Preferences of Broadcast Listeners," *PIRE* 33, no. 9 (September 1945): 571-81; also in *Journal of Experimental Psychology* 35, no. 5 (October 1945): 374-90; quote from Eisenberg and Chinn, "Tonal Range Preferred of Listeners: Results of Study Show Liking for Narrow Reproduction," *Broadcasting-Broadcasting Advertising*, September 17, 1945, 30. My thanks to Gary Frost for providing the latter citation.

[129]Walter van B. Roberts, in "Discussion on 'Tonal-Range and Sound-Intensity Preferences of Broadcast Listeners,'" *PIRE* 34, no. 10 (October 1946), 757-58.

[130]Quotes in ibid., 760; see also James Moir to the editor, *PIRE* 35, no. 5 (May 1947), 495, and J. Moir, "Perfect vs. Pleasing Reproduction," *Audio Engineering*, June 1947, 24-27, 41-43, originally in *Electronic Engineering* (U.K.), January 1947.

[131]Howard A. Chinn and Philip Eisenberg, "Influence of Reproducing System on Tonal-Range Preferences," *PIRE* 36, no. 5 (May 1948), 572-77. Chinn and Eisenberg also investigated listener preferences for loudness, relative and absolute, of broadcasts. Their results were less controversial, although the findings that most people preferred an even level of sound and the same level for speech and music were further tipoffs that most people were not interested in the dynamic possibilities of high fidelity: Howard A. Chinn and Philip Eisenberg, "New C.B.S. Program Transmission Standards," *PIRE* 35 no. 12 (December 1947), 1547-55.

[132]This argument arises most often in relation to Armstrong's efforts to promote FM radio before and after World War II. Susan Douglas follows this line most recently in *Listening In*, 262-63. See also Read and Welch, 338, for their understanding of the international corporate agreements behind the suppression of British-developed phonographs in the U. S. after World War II.

[133]See Walter van B. Roberts, in "Discussion on " 758, and Eisenberg and Chinn, 759, in "Discussion on 'Tonal-Range.'" Roberts worked at Palmer Physical Laboratory at Princeton University and presumably knew about Olson's research.

[134]Harry F. Olson, "Subjective Frequency Tests," RCA Laboratories Report Z-2, January 17, 1944, pp. 1-2, Hagley 2069, series XXII: "'Z' Reports."

[135]Ibid., 2-4; quote, 4.

[136]"Radio Laboratory Tests Show Taste for Natural Music Tones," [Philadelphia *Inquirer*? May 10, 1947?], clipping in back of patent disclosure binder: "Reprints of Papers by Harry F. Olson," Olson Papers.

[137]Harry F. Olson, "Frequency Range Preference for Speech and Music," *Journal of the Acoustic Society of America* 19 no. 4 (July 1947): 549-55, published also in RCA's *Broadcast News* 46 (September 1947), 28-32. See also Olson's oral history at www.ieee.org/organizations/history_center/oral_histories/transcript/olson.html, visited March 1, 2000. For an explanation of the difference in results, see N. M. Haynes, "Factors Influencing Studies of Audio Reproduction Quality," *Audio Engineering*, October 1947, 15-17, 35; and Stephen E. Stuntz, "The Effect of Sound Intensity Level on Judgment of 'Tonal Range' and 'Volume Level,'" *Audio Engineering*, June 1951, 17-19, 26. Carson et al., "A Record Changer," 179, credit Olson for their refinements to the 45's design.

[138]"What Constitutes High Fidelity?" *Audio Engineering*, December 1948, 8, 36-38, quote on 8.

[139]Ibid., 36-38.

[140][John H. Potts], "Transients: High Fidelity," Audio Engineering, May 1947, 4; "Report on Dr. Harry F. Olson's Listener Preference Tests," *Audio Engineering*, June 1947, 27, 43-44; "Today's Radio and Television: Not What They Ought to Be," *Consumers' Research Bulletin*, February 1948, 24-25; "The Average Listener," *Radio & Television News*, May 1949, 153-54.

[141]C. J. LeBel, "Psycho-Acoustic Aspects of Higher Quality Reproduction," *Audio Engineering*, January 1949, 9-11, 32-4, quote on 34.

[142]"Golden Ears and Tin Ears," *Newsweek*, August 11, 1947, 82.

[143]Harry F. Olson, "Wide Range Reproducers and the Tanglewood Demonstrations," PEM-79, August 18, 1947, 9, 10, Sarnoff Corporation Library. This should also be available in Hagley 2069, series III: "Princeton Lab technical reports."

[144]Ibid., 10-11.

[145]Ibid., 4-7, 8; Harry F. Olson, "Audio Noise Reduction," *Electronics*, December 1947, 118-22; and Edward Dickey, ed., "National Electronics Conference: Technical Papers," *RCA Laboratories Division News*, December 1947, 2, which notes Scott's vague terminology and lack of data. For coverage of Scott's invention, see "Hasn't Scratched Yet," *Newsweek*, January 6, 1947, 65; "Shorts and Faces: Music Without Noise," *Fortune*, June 1947, 148-50; H. H. Scott, "Dynamic Suppression of Phonograph Record Noise," *Electronics*, December 1946, 92-95; H. H. Scott, "Dynamic Noise Suppressor," *Electronics*, December 1947, 96-101; Edward Tatnall Canby, "Record Revue," *Audio Engineering*, May 1948, 24, 38-41.

[146]Irving Kolodin for the New York *Sun*, quoted in Olson, "Wide Range Reproducers," 14.

[147]"Golden Ears and Tin Ears," 82; "New Custom-Built Series of Instruments Introduced by RCA Victor Division," *RCA Laboratories Division News*, September 1947, 13; "Introducing the 'Berkshire,'" *Radio Age* 7, no. 1 (October 1947): 22-23.

[148]H. E. Roys, "Improvement in the Quality of Phonograph Reproduction," RCA Victor TR-1006, December 18, 1947, quote on p. 9, Hagley 2069, series XXI: "RCA technical reports," box 121; H. E. Roys, "Intermodulation Distortion Analysis as Applied to Disk Recording and Reproducing Equipment," *PIRE* 35 (October 1947): 1149-52; Murlan S. Corrington, "Tracing Distortion in Phonograph Records," *RCA Review* 10, no. 12 (1949): 241-53; H. F. Olson, "Preliminary Program Phonograph Record Research," witnessed May 3, 1948, 7-8, in Herbert Belar's RCA Laboratories notebook P-4613, Sarnoff Corporation Library. One company's lacquer masters showed three to four percent IMD, compared to ten to twenty percent on the records: LeBel, "Psycho-Acoustic Aspects," 11. Carson et al., "A Record Changer,"179, implicitly admit sound quality had not been a mandate in 1939 by stating only that high fidelity had been "an engineering objective for some time." They cite Roys's work on 180.

[149]Edward Tatnall Canby, "Record Revue: Classical Records," *Audio Engineering*, February 1948, 32, 43-45, and March 1948, 32, 38-40, quote on 32; "What Constitutes High Fidelity," 36.

[150]Hartwell, "The Man Who Sells," 102-103. It is not clear if Folsom identified this market independently.

[151]"RCA Executives Promoted," *Radio Age* 5, no. 2 (January 1946) 22; "Case Report on RCA Victor Records," 11-14, 30-40, and Exhibits D and J, "RCA Victor Advertising Outlay, Popular vs. Classical" and "JWT Personnel Handling RCA Victor Records," 100 and 108, respectively. See also the Columbia ads in the D'Arcy Masius Benton & Bowles collection, box 44, folder 61, Hartman Center; all of those retained for the postwar period promote classical recordings.

[152]See Robert Jourdain, *Music, the Brain, and Ecstasy: How Music Captures Our Imagination* (New York: 1997), Chapter 8, "To Listening . . .," especially 245-49, for a detailed analysis of the difference between hearing and listening.

[153]In late 1947, the *SRL* expanded Canby's column into a monthly section edited by Irving Kolodin, a move that attracted the notice of *Newsweek*: "Music to Read About," September 1, 1947.

[154]Canby, "The New Recordings," *SRL*, January 12, 1946, 29.

[155]Quote from Edward Tatnall Canby, "The New Recordings: Some Comparisons," *SRL*, September 14, 1946, 39; idem, "The New Recordings," *SRL*, January 12, 1946, 29, and February 9, 1946, 41; idem, "Record Revue," *Audio Engineering*, September 1947, 29.

[156]Edward Tatnall Canby, "The New Recordings," *SRL*, April 13, 1946, 92; idem, "The New Recordings: Asch, Disc, Union," *SRL*, May 25, 1946, 45.

[157]Canby, "The New Recordings," *SRL*, January 26, 1946, 42, and February 6, 1946, 26.

[158]Edward Tatnall Canby, "The New Releases: Columbia Releases," *SRL*, June 29, 1946, 45. Michael Gray, in "A History of the Sound," displays an ambivalence about the label's 78s. On the one hand, Columbia had a "reputation for less than exceptional sound that dogged it throughout the 78 rpm era": 31; on the other, its records "sounded better than any in the classical industry, including Decca's FFRR 78s.": 32.

[159]Edward Tatnall Canby, "The New Recordings," *SRL*, January 12, 1946, 29.

[160]Quote from Edward Tatnall Canby, "The New Recordings," *SRL*, January 5, 1946, 22, emphasis in the original; idem, "The New Recordings," *Audio Engineering*, April 13, 1946, 92.

[161]In a random survey of Canby's surface ratings between January and November 1946, Canby gave Columbia releases fifteen A's to Victor's eight. On the other hand, 79% of Victor's recordings received an A- or better compared to 81% of Columbia's. Quote from "The Discographer" [Edwin Tatnall Canby], "The New Recordings," *SRL*, January 12, 1946, 29; Canby's review of Toscanini and the NBC Symphony Orchestra's "Romeo and Juliette," opus 17 by Hector Berliosz, Victor DM 1160 (3) or Victor DV (7) (3 plastic), *Audio Engineering*, March 1948, 40-41. By June 1948, Columbia and a host of smaller labels had switched to Vinylite for their 78s: full-page ad for Bakelite Corporation's Vinylite, *SRL*, June 26, 1948, 45.

[162]See Canby's review of Pittsburgh Symphony, "Symphony no. 40 in G. Minor" by W. A. Mozart, Columbia MMV 727 (3 plastic), *Audio Engineering*, May 1948, 41.

[163]First quote from Canby's review of the Metropolitan Opera Company's "Hansel and Gretel," by Humperdinck, Columbia M OP 26 (2 vols.), *Audio Engineering*, November 1949, 46; second quote from idem, "Record Revue," *Audio Engineering*, November 1949, 34.

[164] "Best of the Year—1949," *SRL*, December 31, 1949, 45.

[165]Edward Tatnall Canby, "The New Recordings: How to Relax," *SRL*, August 10, 1946, 29. Columbia stopped issuing classical albums for manual changing in 1947: "Editorial Notes," *ARG*, October 1947, 33-34.

[166]Canby, "The New Recordings: How to Relax," 29.

[167]Canby, "The New Recordings: Record Changer and Other Matters," *SRL*, August 24, 1946, 31.

[168]Ibid.

[169]The story of the LP's development is based on the recollections of various participants. CBS assigned credit to Peter Goldmark, who embroidered the story in his autobiography with Lee Edson, *Maverick Inventor: My Turbulent Years at CBS* (New York: 1973). Goldmark, 130, described Wallerstein as "an assertive fellow, full of self-importance," who patronized the CBS research director. C. A. Schicke dedicated *Revolution in Sound: A Biography of the Recording Industry*, (Boston: 1974), to Wallerstein and assigns him the primary role in negotiations with RCA. Wallerstein's supporters give him more credit for supporting the project than does Frank Stanton, then president of CBS, in his oral history. In the last two years, various retirees have publicized the engineering effort of William S. Bachman. The following narrative attempts to interpolate these accounts within the framework of CBS's place in the communications industry.

[170]My dating is based on Sally Bedell Smith, *In All His Glory*, 281, and the premise that the disc Goldmark donated to the Smithsonian Institution was his first attempt at longplaying record; a photograph of it appears in Cynthia Hoover, *Music Machines—American Style* (Washington, D.C.: 1971), 118. See also Goldmark, 126-27, and "The Reminiscences of Frank Stanton [part 2]," (1991), pp. 104-105, in the Oral History Collection of Columbia University. My thanks to Dr. Stanton and Mary Marshall Clark of the Oral History Research Office for allowing me to read the final draft of the interview. Goldmark, 131, dismisses Hunt and Pierce's work (see Chapter Five, section xvii, 348-49), along with that of Bell Labs as of minor importance. See John Alvin Pierce, "The Phonograph's Forgotten Heros," *Audio*, March 1991, 46, for the view from Hunt's laboratory at Harvard.

[171]Goldmark, 132-33; Smith, 281. CBS actually caught up with NBC in number of affiliates when the latter had to sell its Blue network in 1942-45. The number did not reflect the size of the audience, however, and Paley continued to promote CBS's reputation for quality.

[172]William S. Bachman notes that "CBS was not unaware of the fine-groove recording work proceeding at RCA" in "The LP and the Single," *JAES* 25, no. 10/11 (October/November 1977), 823. For Snepvangers, see "Stylus," "Characteristics of the New 45 RPM Record," *Audio Engineering*, March 1949, 47; "1960 AES Awards: the Emile Berliner Award," *Journal of the Audio Engineering Society* 9, no. 1 (January 1961): 72. Snepvangers worked at RCA Victor until at least November 1944: see patent no. 2,426,061. He left Camden because his wife preferred New York: Murlan S. Corrington to the writer, June 6, 1995, p. 1. Wallerstein, in "Creating the LP Record," 57, called Snepvangers "a key factor in the success of our plans."

[173]Wallerstein, "Creating the LP Record," 58; Stanton OH [part 14], 734-36; Stanton to the writer, August 11, 1997, p. 1; Schicke, 96, 122-23; Goldmark, 133-37; Pierce, "The Phonograph's Forgotten Heros," 46.

[174]"Stylus," "Characteristics of the New 45 RPM Record," *Audio Engineering*, March 1949, 47; W. S. Bachman, "Phonograph Reproducer Design," *AIEE Transactions* 65 (March 1946): 159-62.

[175]Schicke, 122; Wallerstein, "Creating the LP Record," 58; Goldmark, 146; Frank Stanton to the writer, August 11, 1997, p. 1; transcription of "National Music Arts Presents: 50th Anniversary of the Long-Play 33 1/3 Record," National Academy of Sciences, Washington, D. C., January 18, 1998, p. 2, courtesy of Susan Schmidt Horning, Case Western Reserve University.

[176]Wallerstein, "Creating the LP Record," 58; for more on Paley's resistance to capital investment, see Stanton OH [part 7], 374-77.

[177]Pierce, "The Phonograph's Forgotten Heros," 46. Pierce separates Columbia's offer from the outright theft by other companies in the industry, but I infer that Bachman and Snepvangers used their patents anyway because of their basic nature to microgroove reproduction.

[178]Robert Angus, "Why do Records Have to be Black Anyway?" *High Fidelity*, July 1974, 68. My thanks to Dr. Robert Friedel, University of Maryland, College Park, for recalling a similar article by Angus, "Whatever Happened to Pure Virgin Vinyl? Why Records are Black," *Forecast!*, February 1975, 60, and to Carlene Stephens, National Museum of American History, for finding this one. Angus's interviewees in the industry responded to the question from a contemporary perspective marked by their efforts to make records during the Arab oil embargo. Polyvinyl acetate, the copolymer in the record compound, also helped reduce groove noise. Because of the difficulty in obtaining it in 1973-74, manufacturers resorted to using more carbon black to compensate: Gene Lees, "The Vinyl Shortage: Does It Mean Poorer and Fewer Records?" *High Fidelity*, July 1974, 70.

[179]Lieberson, "A 33 1/3 Revolution in Recording," 41; William S. Bachman, "From Transcription Disc to LP," *High Fidelity*, April 1976, 59; Hobson, "The Classic Interview;" transcription of "National Music Arts Presents," pp. 4-5. To be sure, Columbia issued shorter classical pieces on ten-inch LPs, and it and other labels began pressing twelve-inch popular LPs in short order.

[180]Goldmark, 138-40; Schicke, 122-23; Smith, 282.

[181]Smith, 282; Stanton OH [part 14], 737-38; Bachman, "The LP and the Single," 823. Schicke, 119-22, credits Wallerstein rather than Stanton for these negotiations.

[182]Schicke, 120-21; Smith, 282; Bachman, 823; Kenneth Bilby, *The General: David Sarnoff and the rise of the communications industry* (New York; 1986), 262; quotes from *Color Television: Testimony of Brig. General David Sarnoff, Chairman of the Board, Radio Corporation of America, before the Federal Communications Commission*, (Washington, D. C., May 3 and 4, 1950), 256, DSL.

[183]Schicke, 123; Hobson, "The Classic Interview."

[184]In one example of the diffusion of the standard histories into other fields, one business writer used this format battle as an example of a strategy taught at Harvard Business School. Lawrence Shames concluded that after Columbia knocked out the market for 78s with the LP, RCA "ad prac'ed" or overcame the innovation by inventing a record playable only on their turntables. See Shames, *The Big Time: The Harvard Business School's Most Successful Class and How It Shaped America* (New York: 1986), 36.

[185]For RCA's turnaround at Atlantic City, see Warren Rex Isom, "A Wonderful Invention," (Bloomington, IN: videotape produced by RCA Records through Indiana University TV, 1985). Courtesy of Mr. Isom, who rose from sound reproduction engineer to chief engineer of RCA Records between 1944 and 1976. For the other aspects of the debate, see "Disc Reproducer Considerations," June 9, 1948, and the entry for August 31, 1948, in Belar's lab notebook P-4612, Sarnoff Corporation library.

[186]Carson et al., "A Record Changer," 186-87; "Stylus," "Characteristics," 46; *Color Television*, 257.

[187]The evidence for RCA's flirtation with a large diameter 45 is fragmentary: "Deny RCA Victor Will Manufacture Long-Playing Disc," *Retailing Daily*, July 7, 1948, 63, repeated July 8, 45; H. Belar, A. R. Morgan, and H. F. Olson, "The Super 45," November 6, 1950, p. 1, Olson Papers, file: "The Super 45," Sarnoff Corporation library; Phil Wooley to the writer, September 11, 1995, p. 2, and October 29, 1995, p. 3. Wooley joined RCA early in 1948 to work on audio in the Home Instruments Department.

[188]The board of directors elected Folsom president December 3, 1948, to take effect January 1, 1949, on Sarnoff's recommendation: "Changes in RCA Management," *Radio Age* 9, no. 2 (January 1949): 7.

[189]"RCA Gets Into the Battle of the Discs," *Business Week*, January 22, 1949, 82-85; "The General's Biggest Battle," *Newsweek*, December 5, 1949, 62-66; "Business Roundup: Phonograph Record Boom," *Fortune*, January 1950, 15; Malcolm Forbes, "Fact and Comment: The Bigger They Are—," *Forbes*, February 15, 1950, 11; RCA's rejoinder, "RCA Officials Hit Back at Forbes Criticism," *Forbes*, March 15, 1950, 17-18; "At the End of the Rainbow," *Time*, December 4, 1950, 52; C.B.S. Steals the Show," *Fortune*, July 1953, in *RCA: The Years 1930-1978*, 35-40.

[190]While the standard biographers and historians of RCA discuss Sarnoff's background and his ego, none of them link these factors explicitly to his sense of personal and corporate destiny. See Bilby; Carl Dreher, *Sarnoff: An American Success* (New York, 1977); Graham, 30-75; Lewis, 89-118; Eugene Lyons, *David Sarnoff* (New York, 1966); Sobel, 42-45.

[191]Bilby, 262-63. It should be noted that NBC had already lured away Ozzie and Harriet and two other programs from CBS: "NBC Moves Ahead of CBS in Fall Program Shuffle," *Advertising Age*, July 19, 1948, 49. For Sarnoff's opinion, see his testimony in *Color Television*, and his memo of May 7, 1951, "David Sarnoff Papers," vol. 3, "Color" binder, DSL.

[192]First and last quotes from [Peter Hugh Reed], "Columbia's Long-Playing Record," *ARG*, July 1948, 330; middle two quotes from idem, "Editorial Notes," *ARG*, August 1948, 361; for the osmium stylus, see idem, "Editorial Notes," *ARG*, December 1948, 97.

[193]Quotes from [Reed], "Editorial Notes," August 1948, 368-9.

[194]Frank Stanton to the writer, August 11, 1997, p. 2; "Columbia's Challenge," 82; quotes from "Video Overshadows Radios at Closing Chicago Exhibition," *Retailing Daily*, July 16, 1948, 19; "Bendix Announces Direct Sales Plan at Furniture Show," *Advertising Age*, July 12, 1948, 67. Philco's announcement came after one for Congoleum-Nairn's new asphalt linoleum tile. For more on the crystal-based pickup, see "Modern Crystal Phono Pickups," 29.

[195] "Deny RCA Victor Will Manufacture Long-Playing Disc," *Retailing Daily*, July 7, 1948, 63, repeated July 8, 45; "Phila. Stores Score LP Disc Sales Policy," *Retailing Daily*, July 21, 1948, 35; "Poll Shows Sharp Contrasts in Long Playing Record Setup," *Retailing Daily*, July 23, 1948, 19; "New Coercion Charges Hurled in Record Sales controversy," *Retailing Daily*, July 28, 1948, 30.

[196]Ad copy quoted in "Columbia Records Break Campaign for LP Discs," *Advertising Age*, August 2, 1948, 6; last quote from "Rough Proofs" commentary, *Advertising Age*, August 9, 1948, 1. For more on the solution to the problem of record storage, see "The Ideal Record Cabinet," *ARG*, June 1948, 299-301.

[197]C. G. McProud, "Columbia LP Microgroove Records," *Audio Engineering*, August 1948, 24.

[198][Donald G. Fink], "Transcription Recordings for the Home," *Electronics*, September 1948, 86-87.

[199]M. Harvey Gernsback, "Microgroove Phonograph Records," *Radio-Electronics*, October 1948, 30-31; see also John B. Ledbetter, "Microgrooves Mean More Money," *Radio & Television News*, January 1949, 54-55, 108-10, for the implications for the service industry.

[200]C. G. Burke, "Report on LP: the Question of Wear," *SRL*, February 26, 1949, 56; Burke changed his mind about the efficacy of osmium points and the durability of microgrooves in "Report on LP: Random Notes," *SRL*, January 28, 1950, 70. For reader response, see "Letters to the 'Recordings' Editor, *SRL*, April 30, 1949, 62.

[201] "45 RPM Windows, 33 1/3 RPM Displays," in Harrison (1949), 90-91.

[202] "The New Haven Record Dealer Who Knows How," in Harrison (1949), 41.

[203]"Executive Opinion: 'The Influx of Independents," *Business Week*, October 15, 1949, 34-35; "Two Sides to Three-Speed Record Mixup," *Business Week*, April 8, 1950, 80-81; "New Market for New Discs," *Business Week*, April 8, 1950, 82-83; "Case Report on RCA Victor Records," JWT Archives, 66-67.

[204] "Executive Opinion," 34.

[205]For phonograph production, see "Historical Files: Firsts: First 45 rpm attachment—production totals," Hagley 2069, box 291. These files have since been reorganized in the Historian's series. "Pee Wee" was a re-issue of a 1947 78 and is displayed at the Thomson History Center, Thomson/RCA Consumer Electronics, Indianapolis. For the four-disc press and union protest, Phil Wooley to the writer, October 29, 1995, p. 2.

[206]"Record Mixup," *Time*, December 27, 1948, 52.

[207]Edward Tatnall Canby, "Two Standards—or Three?" *SRL*, October 30, 1948, 60; Irving Kolodin, "Impressions," *SRL*, October 30, 1948, 45.

[208] "New Record Scramble?" *Newsweek*, November 8, 1948, 85-86; "Record Mixup," *Time*, December 27, 1948, 52; "One for Harry," *Time*, December 27, 1948, 56. For more on Time's anti-union stance, see the cover story on Petrillo, "The Pied Piper of Chi," *Time*, January 26, 1948, 18-22.

[209]"RCA Victor, Columbia Wage Battle Over New-Type, Long-Playing Discs," newspaper clipping of United Press story, January 10, 1949, courtesy Claudia Depkin, BMG Entertainment Archives, New York; "Out of the Groove," 82-84.

[210] [Peter Hugh Reed], "Recording—*Quo Vadis*?" *ARG*, January 1949, 130, 134-35.

[211]Ibid., 130; "Record Mixup," 52.

[212]James W. Murray, "Standardization and RCA," *SRL*, January 29, 1949, 49-50.

[213]Irving Kolodin, "Impressions," and "Letters to the RECORDINGS Editor," *SRL*, January 29, 1949, 47, 66; [Peter Hugh Reed], "New Paths in Recording," *ARG*, February 1949, 162.

[214]Boston Pops on RCA Victor 49-0100; Crudup on 50-0000. My thanks to Claudia Depkin, BMG archivist, for providing the release information.

[215]Irving Kolodin, "45 RPM—First Phase," *SRL*, March 26, 1949, 56.

[216]A less well-informed article endorsing the 45 is Tom Gootée, "The 'New Look' in Popular Records," *Radio & Television News*, March 1949, 41, 96-97.

[217]"Stylus," "Characteristics of the New 45 RPM Record," 6. Given his pseudonym and familiarity with both companies' technology, Stylus might have been W. H. Rose of Frank L. Capps and Company, which manufactured styli for the recording industry. Before World War II, Rose lunched occasionally with friends from NBC's Recording Division and Columbia Records: Robert J. Callen, "Hollywood Sapphire Group," *Audio Engineering*, January 1948, 17.

[218]First quote from "Letters: Comment on Comments," *Audio Engineering*, May 1949, 4, 41, in response to "Comments on 45 Report," Leonard Carduner to the editor, *Audio Engineering*, May 1949, 4; sec-

ond quote from "Stylus," "Characteristics of the New 45 RPM Record," 47; recording study in "Stylus," "The Cutting Stylus [sic] Problem in Microgroove Recording," *Audio Engineering*, April 1949, 26-28.

[219][Peter Hugh Reed], "Preamble On the New Records: Editorial Notes," *ARG*, March 1949, 194.

[220]Peter C. Goldmark, René Snepvangers, and William S. Bachman, "The Columbia Long-Playing Microgroove Recording System," *PIRE* 37, no. 8 (August 1949): 923; Charles Hobbs to the writer, June 29, 1996, p. 1, and August 12, 1996, p. 5.

[221]Edward Tatnall Canby, "Record Revue," *Audio Engineering*, March 1949, 30; see also "Letters: LP Reactions, Richard H. Dorf to the editor, *Audio Engineering*, April 1949, 6, 46; [Peter Hugh Reed], "Recording—*Quo Vadis?*" *ARG*, January 1949, 130; "The Three Types of Phonograph Records—Which to Buy?" *Consumers' Research Bulletin*, November 1949, 17. The writer of the latter article was a professional engineer who also taught music.

[222]Joseph G. Wilson, RCA Victor executive vice-president, died in June 1950: "Buck, Elliott Inherit Most of Big Job at RCA Victor Left by Death of Wilson," *Billboard*, June 24, 1950, 4; "Case Report on RCA Victor Records," JWT Archives, 61-65.

[223]Hartwell, "The Man Who Sells," 101-102; "Case Report on RCA Victor Records," 76, formula reproduced as Exhibit I, 107. Of Folsom, the anonymous agency historian wrote, "It is not the job of a Case Report to dip into psychoanalysis, but it is no secret that Folsom was hard to get along with": 50.

[224]Calculated from Sterling and Haight, Tables 100-B, 650-B, and 660-B on pp. 4, 357, and 363 respectively.

[225]In 1960, dealers made forty percent of their phonograph sales in the fourth quarter and twenty percent in the thirty days before Christmas: "RCA-General," February 8, 1961, part (4), in JWT Archives, Review Board box 25, folder: "Meetings: RCA Corporation General Summaries, 1961-1966," Hartman Center. These percentages do not differ significantly from those in 1920.

[226]Paula Dranov, *Inside the Music Publishing Industry*, (White Plains, NY: 1980), Figure 6.1, "Record Distribution in the 1930s to early 1940s," 112; The Editors, "Opportunities! . . . in Record Shops," *Forbes*, August 1, 1947, 18. See also the articles on record store design and display and the relationship to self-service in [Neil F. Harrison, ed.], *Record Retailing Yearbook* (New York: 1948 ed.), especially "The Blueprint for Tomorrow's Record Store Puts Customer Convenience and Comfort First, 44-48; "The 'Shop of Tomorrow'—Streamlined, Purposeful Elegance," 53-54; and Raymond L. Green, "Singles, Self-Service Style, Make Their Sales Soar," 57-58.

[227]Mike Ranalli to the writer, April 10, 1995, pp. 1-2. Ranalli was in charge of the 45's sales program for the Home Instruments Department. Similar pressure explained successful sales of RCA radios during 1948's slump: "R.C.A.'s Television," *Fortune*, September 1948, in *RCA: The Years 1930-1978*, 31.

[228]Ranalli to the writer, p. 3; [Reed], 'Preamble on the New Records," 193; "45 RPM Windows, 33 1/3 RPM Displays," 90-91; Jack M. Williams, "45 RPM—An Adventure in Unique Merchandising for the Imaginative Dealer," and quotes from John D. Benedito, "What Plastics Mean to the Recording Industry," both in Harrison (1949): 16-17 and 25 respectively; see ads reproduced in "Case Report on RCA Victor Records," JWT Archives, after p. 63.

[229]Warren Rex Isom, "Record Materials Part II: Evolution of the Disc Talking Machine," *JAES* 25, no. 10/11 (October/November 1977), 722. The only people I have met who recalled the 45s' debut clearly also recalled the impression of cheapness: Robert Heinz to the writer, April 10, 1998, and Charles Rhodes to the writer, January 7, 2000. See Jeffrey Meikle, *American Plastic: A Cultural History* (New Brunswick, NJ: 1995), for the best explanation of the place of plastic in American culture.

[230]The issue was confused first by the distinction between the time of the changer's operation and the time from music on disc to music on the next. RCA also simplified the changer mechanism in 1950, which had an eight-second cycle compared to five for the original model. See Carson et al., "A Record Changer," 174, 179; "New RCA-Victor '45's'—results of CU's first listening tests," *Consumer Reports*, May 1949, 197; "Comments on 45 Report," 4; "Stylus," "Comment on Comments," 4, 41; "The Three Types of Phonograph Records—Which to Buy?" *Consumers' Research Bulletin*, November 1949, 17; Goldmark, 142-3. In 1931, the Electromatic and Electra record changers had change specifications of five and seven seconds respectively: Irby, "Developments," 586.

[231]Toscanini Legacy, folder M56 B-2, New York Public Library for the Performing Art—Music Division, New York City. Despite the use of the Legacy's database, I could find no reference to Toscanini's opinion of the 45.

[232]See the happy couple enjoying Edward R. Murrow's "I Can Hear It Now," vol. II, in the full-page ad in *Life*, January 23, 1950, 5.

[233] "Record Dither," *Time*, September 5, 1949, 52; Edward Wallerstein, "A Review of Columbia's New Long Playing LP Microgroove Records," in Harrison, (1949): 15.

[234] "Record Dither," 52; "The General's Biggest Battle," *Newsweek*, December 5, 1949, 62; Wallerstein, "Creating the LP," 61.

[235]"Fight Between the Record Makers—the innocent customer takes a beating," *Changing Times: The Kiplinger Magazine*, August 1949, 39; "The Three Types of Phonograph Records," 15-19, quotes from 18, 19.

[236] "Fight Between the Record Makers," 39.

[237]In 1949, consumers' disposable income declined .2% in constant dollars: Sterling and Haight, Table 300-C, 114. See also "Business Roundup: 1949 in Retrospect," *Fortune*, January 1950, 11.

[238]Anderson et al, "Hit Record Trends," Table 1, 33; "Columbia Seen Cutting Retail Price to 60¢, Setting Trend," *Variety*, March 23, 1949, 37; "CRC Into 49¢ (inc. tax) Disks via Harmony Label," *Variety*, June 8, 1949, 45; "RCA 50%-Off Clicks Big, But Also Helps Other Regular-price Sales," *Variety*, June 16, 1949, 39; "RCA Decides on 49¢ Price for Cheaper Label; Revive Bluebird Tag," *Variety*, July 6, 1949, 39; Traiman, "The Record Industry," 786, but see Sterling and Haight for industry sales in constant dollars, Table 350-A, 190, and Table 650-B, 357, for the decline in sales volume of phonographs.

[239]"It's Music, Music, Music," *Business Week*, July 22, 1950, 32. See also the rise in number of entries related to music, as well as phonographs, in the *Readers' Guide to Periodical Literature*; and *The Music Index 1949: Annual Cumulation*, (Detroit: 1949). Florence Kretzschmar began *The Index* in January 1949, citing "revolutionary developments in television, microgroove recording, and electronics" as stimulating "new audiences and a greatly increased public interest" (1). The annual volumes jumped in size from 308 to 505 pages in two years.

[240]"Case Report on RCA Victor Records," JWT Archives, 56-58.

[241]Ibid., 57, 60-61.

[242]Ibid., 57.

[243]"The Public Has Chosen 45," (RCA Victor: 1950), Hagley 2069, Historian's series II, box 3, folder 5: "Mixed Product - Trade Catalogs and Instruction Manuals, 1928-1950."

[244]Michael Gray, "Record Reissues—An American Perspective; An Interview with John Pfeiffer, RCA Records," *ARSC Journal* 8, no. 2/3 (1976), 6; "Jack Pfeiffer's Corner: The Audio Interview," p. 2, http://www.classicrecs.com/jackint2.htm, visited February 17, 1999; "Case Report on RCA Victor Records," JWT Archives, 59-60; "The World's Greatest Artists Applaud RCA Victor's 45 rpm System," full-page advertisement in *SRL*, October 29, 1949, 49; see also the ad proofs in the RCA Victor advertising collection, Camden County Historical Society, Camden, New Jersey.

[245]Full-page ad, "RCA Victor Announces . . . ," *SRL*, January 28, 1950, 65, which repeated the message of a booklet, "Facts About Records," (RCA Victor: c. 1950), courtesy Claudia Depkin, BMG Entertainment Archives.

[246]Irving Kolodin, "Impressions," *SRL*, January 28, 1950, 55.

[247]Quote from [Irving Kolodin], "RCA Goes to LP," *SRL*, February 25, 1950, 60.

[248]Quote from C. G. Burke, "Report on LP: Random Notes," *SRL*, January 28, 1950, 70; idem, "78-45=33," *SRL*, April 29, 1950, 58-59.

[249]Eyewitness Television was an advertising term used by RCA Victor when television was introduced after World War II.

[250]By using connectors between the component parts like motor, amplifier, and speaker, the technician could take everything apart easily to facilitate repair.

[251]For more information regarding spindle designs, refer to rp-168 changer in Chapter 14.

[252]The superheterodyne was a quantum leap in radio improvement credited to Edwin Howard Armstrong.

[253]Push-pull is a type of audio output circuit that uses tw tical vacuum tubes, each one amplifying half of the signa circuit sounds much better than the single tube or single- audio circuit.

[254]*RCA 45EY2 Phonograph Restoration with Paul Child* a two-hour VHS video that shows how to rebuild the record ch rebuild the amplifier, and restore the Bakelite cabinet.

Bibliography

Bachman, W. S. "Stylus, Characteristics of the New 45 RPM Record." *Audio Engineering*, March 1949.

Barnum, Frederick O. III. *"His Master's Voice" in America*. Camden, NJ: General Electric Company, 1991.

Baumbach, Robert W. *Look for the Dog*. Woodland Hills: Stationary X-Press, 1996.

Bilby, Kenneth. *The General: David Sarnoff and the Rise of the Communications Industry*. New York: Harper Collins, 1986.

Carson, Benjamin, Alexander Burt, and Hillel Reiskind. "A Record Changer and Record of Complementary Design." *RCA Review* 10, June 1949.

Carson et al. "A Record Changer." RCA Victor internal memorandum, Princeton, New Jersey.

Childress, Paul. "Restoring the 45EY2 Phonograph" *Video*, 1998.

Dreher, Carl. *Sarnoff: An American Success*. New York: Quadrangle, 1977.

Ecklund, Eugene. *Repairing Record Changers*. McGraw-Hill Book Company, 1955.

Fabrizio, Timothy C., and George F. Paul. *Phonographs with Flair: A Century of Style in Sound Reproduction*. Atglen, PA: Schiffer Publishing Ltd., 2001.

Gelatt, Roland. *The Fabulous Phonograph: From Edison to Stereo*. New York: Appleton-Century, 1965.

Goldmark, Peter. *Maverick Inventor: My Turbulent Years at CBS*. New York: Saturday Review Press, 1973.

Lewis, Thomas. *Empire of the Air: The Men who Made Radio*. HarperPerrinnial, 1993.

Lyons, Eugene. *David Sarnoff*. New York: Harper & Row, 1966.

Magoun, Alexander Boyden. *Shaping the Sound of Music, The Evolution of the Phonograph Record, 1877-1950*. College Park, Maryland: Dissertation submitted to the Graduate School of the University of Maryland, 2000.

Marco, Guy A. and Frank Andrews. *Encyclopedia of Recc Sound in the United States*. New York: Garland Publi Inc., 1993.

Martland, Peter. *Since Records Began; EMI: The First Hur Years*. Portland, Oregon: Timber Press, 1997.

Michigan Antique Phonograph Society. *In the Groove: F Five at Fifty*. February, 1999.

Metz, Robert. *CBS: Reflections in a Bloodshot Eye*. Chic Playboy, 1975.

Moore, Jerrold Northrop. *A Matter of Records*. New Taplinger, 1976.

Paper, Lewis J. *Empire: William S. Paley and the Makin CBS*. New York: St. Martin's Press, 1987.

Read, Oliver, and Walter L. Welch. *From Tin Foil to Stereo: l lution of the Phonograph*. Indianapolis: Howard W. San Co., Inc., 1959.

Reiss, Eric L. *The Compleat Talking Machine*. Chandler, Sonoran Publishing, 1998.

Sams, Thomas W. *Automatic Record Changer Service Man Volumes 3, 4, and 5*. Indianapolis, Indiana: Howard W. Sa & Co., Inc., 1950.

Sarnoff, David. *Looking Ahead: The Papers of David Sarn* New York: McGraw-Hill, 1968.

Slater, Robert. *This ... is CBS: A Chronicle of 60 Yea* Englewood Cliffs, New Jersey: Prentice-Hall, 1988.

Smithsonian Institute. *History of Music Machines*. Drake Pu lishers, 1975.

Sobel, Robert. *RCA*. New York: Stein & Day, 1986.